Landforms
and Landscapes

The **Foundations of Earth Science Series** presents aspects of earth science in convenient volumes suitable for use by students who have had no previous courses in the area. Each volume presents a fundamental framework of the subject in some detail, discussing wherever appropriate, new concepts and current philosophies regarding man's scientific understanding of the earth. Throughout the series, emphasis is placed on environmental considerations and the distribution of the earth's components and resources. Titles in the series are:

David L. Clark—**Fossils, Paleontology and Evolution**

John W. Harbaugh—**Stratigraphy and Geologic Time,** 2nd edition

Theodore G. Mehlin—**Astronomy and the Origin of the Earth,** 2nd edition

Morris Petersen, Keith Rigby, and Lehi F. Hintze—**Historical Geology of North America**

George R. Rumney—**The Geosystem: Dynamic Integration of Land, Sea, and Air**

John S. Sumner—**Geophysics, Geologic Structures, and Tectonics**

Daniel S. Turner—**Applied Earth Science**

Sherwood D. Tuttle—**Landforms and Landscapes,** 2nd edition

Landforms and Landscapes

Second Edition

Sherwood D. Tuttle
University of Iowa

WM. C. BROWN COMPANY PUBLISHERS
Dubuque, Iowa

FOUNDATIONS OF EARTH SCIENCE SERIES

Consulting Editor

Dr. Sherwood D. Tuttle
University of Iowa

Copyright © 1970, 1975 by
Wm. C. Brown Company, Publishers

ISBN 0–697–05008–4

Library of Congress Library Card Number: 78-118886

Printed in the United States of America.

Contents

Preface

To many nonscience majors, the development of ideas in a scientific discipline is the most interesting part of a science course. It is for these students that this book has been written, with the hope that they may find much that is relevant to their vocational and avocational interests. Included here are philosophical concepts from geology, geography, engineering, and other disciplines. Some descriptions and details are given to help students to participate mentally in the analysis and the criticism of concepts, without submerging them in voluminous terminology and calculations. Emphasis has been placed on some of the grand and speculative ideas about the nature and origin of landforms, so that students can know how these theories evolved, what the strong points are, and which ones have flaws. Inconsistencies, controversial and unsolved problems have been sought out because the myth of scientism—that all can be proven—dies hard. Historic viewpoints are reviewed; for to understand a modern position or to reject a well worn hypothesis, one needs an understanding of how past concepts were the products of their times. The influence of new techniques and quantitative procedures is shown through a discussion of their contributions to advances in the earth sciences. An effort is also made to point out limitations of these tools. In a general sense, because it concentrates on landforms and landscapes, this volume is a textbook of beginning geomorphology.

The arrangement and selection of topics and points of view are mine, but I am greatly indebted to many sources—my teachers (especially George W. White and the late Kirk Bryan), my colleagues, and my students. I have tried in all cases to synthesize or summarize the work of others as objectively and accurately as possible, but if there are errors of fact or interpretation, the fault is mine. A special acknowledgment is due my wife Esther for her editorial services during the preparation of both the first and second editions.

In this revised and updated edition of *Landforms and Landscapes*, I want to express appreciation for comments, suggestions, and encouragement I have received from teachers and students who have used the book. Aside from new material and more illustrations (a frequent request), the main change I have made is in the arrangement and sequence of topics. After the introduction, weathering, mass wasting, streams, and stream-eroded landscapes comprise the next three chapters— over one-half the book—because stream processes are the most significant factors in landscape development. Discussion of geomorphic concepts follows because most of them deal with pluvial-fluvial topography. Glaciation is a next logical topic because glaciers, in the main, modify stream-developed landscapes. The final chapter groups together geomorphic agents—wind, waves, ground water, and volcanism—that might be considered the "special effects bunch." Though minor in terms of global accomplishment, they have been of tremendous interest and concern to man throughout history, especially when he has been exposed to their sporadically awesome power.

Sherwood D. Tuttle

Chapter 1

Introduction
to Geomorphology

To many who travel through the countryside or pass over it in an airplane, irregularities of terrain are a jumble of ups and downs in chaotic disarray, but some people—airplane pilots, farmers, hunters, skiers, mountain climbers, infantrymen, and others—look at the landscapes of the earth's surface very carefully. They may be amateur students of landscapes, but all have an urgent need to interpret the variations, large and small, in the topography; their success in this skill may mean accomplishment and a safe return. Professionals who study the earth's surface—the geographers, geologists, and engineers—all have had considerable training and long experience in analyzing the terrain for their particular needs. The purpose of this book is to use some of the methods and insights of both professionals and amateurs to help the curious student gain proficiency in "seeing the lay of the land" so that he, too, can use knowledge and appreciation of terrain in his daily life and work.

An observer perceives land features in different proportions depending on the perspective. From a hilltop, the landscape may be a winding valley at the foot of a gentle ridge; from higher up, with an expanding view, he may see steep mountain slopes, broad level plains, and valleys of many sizes and shapes; from still higher, as if overlooking the world from a space platform, he can see ocean basins and the continents. To bring order and understanding to this great array, he needs a terminology, so that the features can be separated or grouped together descriptively, and a "key" from which their origins can be inferred. *Geomorphology* provides a "language" and a group of philosophical concepts useful in understanding the earth we live on. The study of the configuration of the earth's surface has had a variety of names, the two most common being physiography and geomorphology. The latter designation is the one used more generally at the present time. "Earth-form study" is a literal translation of the Greek roots of the term, with "form" referring to topographic features and not to the geometric form of the entire globe, the study of which is geodesy.

Methods and approaches from both geology and geography are needed in the geomorphic study of landforms and landscapes. *Landforms* are the individual features seen; combinations of features are the *landscapes*. Thus a hill or a single stream valley is a landform, while an area of hills of varying shapes and sizes with streams flowing among them comprises a landscape. An assemblage of landforms makes up the topography, or *terrain*, of an area (Fairbridge, 1968; Thornbury, 1969).

Effects of Geologic Structure on Landscapes

The distribution of major landforms over the surface of the earth is not regular, but is not haphazard, either. Movement of vast plates of crustal material has determined the location of mountain systems. According to theories of plate tectonics, large, rigid units of the earth's less dense crust (several miles thick) move slowly over slightly denser, plastic, subcrustal rock. By this process, continental-sized plates are shifted over the earth's surface. On the edges of moving plates, where the crust is bumping into or being pulled away from the edges of other plates, great mountain systems develop. In a genetic classification of mountains (i.e., fold mountains, fault mountains, volcanic, erosional, and complex mountains), all but erosional mountains owe their location, size, and shape to tectonic activity.

Structure, then, is of fundamental importance in understanding the nature of landforms and landscapes. Geologists use the term *structure* to mean the attitude of the rocks (folded, faulted, etc.), the kind or type of rock (lithology), and the geometric arrangement of rock units and individual formations. Regardless of a particular geomorphic process or climatic regimen dominant in an area, great canyons, mountain valleys, and other elements of "scenery" cannot be formed unless rocks have been structurally moved upward to create mountains or plateaus in the first place. Complex, structurally controlled drainage and topography are mainly the result of erosion working on previously deformed, uplifted, or deposited rocks (fig. 1.1).

EROSION

Some confusion in terminology exists with regard to using a single word to account for most of the changes occurring on the earth's surface,

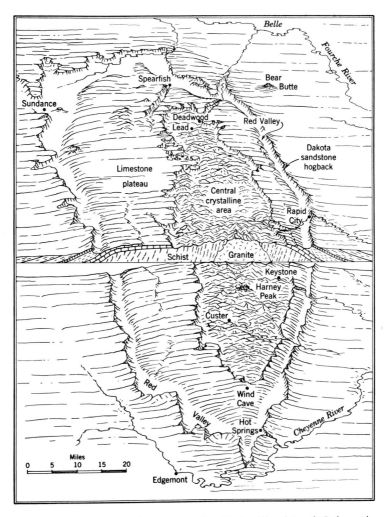

Figure 1.1. The shape and pattern of the Black Hills of South Dakota show erosion working on domed-up sedimentary rocks capping an intruded core of granite. Each up-arched sedimentary unit of Paleozoic or Mesozoic age weathers and erodes differently with the result that distinctive slopes, talus, and drainage develop on each kind of rock. Linear and circular ridges and valleys form on the outcrop patterns of these rocks. In the center of the dome, the more resistant granite forms the highest feature, Harney Peak (7,242 feet). Overall drainage is governed by the geologic structure, with streams flowing outward from the center, then along the outcrop patterns of sedimentary rock units, and eventually crossing more resistant rock in gorges. (From A. N. Strahler, 1969, **Physical Geography,** 3rd ed., John Wiley and Sons. Copyright © 1969.)

but the simplest and most generally accepted term is *erosion*. When we say the Rocky Mountains are undergoing erosion, what is implied by this usage? Essentially, a group of closely related processes are included in the term: (1) *weathering*, the disintegration and decomposition of the rocky materials of the earth's crust; (2) *mass wasting*, the downslope movement of soil and rock; (3) *local erosion*, the entrainment or picking up of material; (4) *transportation* or movement of material by a variety of agents; and (5) *deposition*. Put in equational form we have:

erosion = ± weathering ± mass wasting ± entrainment ± transportation ± deposition.

Note that weathering can occur without movement, and movement (entrainment and transportation) without weathering, although weathering usually goes on before, during, and after transportation of material. The term erosion is used in a more restricted sense when processes of entrainment, such as *stream erosion* or *glacial erosion*, are being discussed. Here erosion denotes the actual picking up of material and its subsequent transportation to a depositional site.

Depending on how they were formed, landforms may be either erosional (i.e., their shape depends on what is left after material has been removed) or depositional, with size and shape being determined by material laid down. Stream valleys are examples of *erosional landforms;* sand dunes and deltas are *depositional landforms*. A few landforms—an outwash plain, for example—are partly erosional and partly depositional.

The principal *geomorphic agents* that bring about changes on the earth's crust are running water, glacial ice, the wind, ground water, and wind-driven water waves. Their activities are referred to as *geomorphic processes*. For example, running water (an agent) does geologic work by stream erosion (a process) (Ruhe, 1975).

GEOMORPHOLOGY AND GEOLOGIC TIME

From historical geology we learn that most of the earth's rocks and structures are much older than the landforms and landscapes of the earth's surface. Most bedrock is more than 60 million years old and the oldest rocks go back about four billion years. A general landmark within the recent history of the earth was the "Great Ice Age." In late Tertiary time, about two or three million years ago, the climates of the earth shifted slowly, precipitation increased, and in some areas glacial ice began to spread over the land surface in vast icecaps or sheets. The earth entered the *Quaternary Period*, which extends to the present. The *Pleistocene Epoch*, which is the main part of the Quaternary, includes the time of the Great Ice Age Post-Pleistocene time is usually called *Recent*, although some experts prefer to call all post-Teritiary time the *Pleistocene* (fig. 1.2).

In the equatorial and low latitude regions of the Southern Hemisphere, some present-day landscapes may have had their beginnings in

Cenozoic Era $\left\{\begin{array}{l}\text{Quaternary Period} \quad \left\{\begin{array}{l}\text{Recent Epoch} \\ (\pm\ 10{,}000\ \text{years}) \\ \\ \text{Pleistocene Epoch} \\ (\pm\ 2\text{-}3\ \text{million years})\end{array}\right. \\ \\ \text{Tertiary Period} \\ (60\ \text{million years})\end{array}\right.$

Mesozoic Era (170 million years)
Paleozoic Era (330 million years)

Precambrian time (several billion years)

Figure 1.2.   Geologic time table, related to the study of landscapes.

the Mesozoic Period; but in the humid temperate areas of the world, because of glaciation and concomitant erosion, the topography is much younger, most of it being Pleistocene in age. In both North America and Eurasia, four separate and distinct ice sheets started, grew, and then melted away. In North America these are named, from oldest to youngest: Nebraskan, Kansan, Illinoian, and Wisconsin. The Wisconsin ice sheet achieved its maximum size between 30,000 to 25,000 years ago. After that time, continental ice retreated or stagnated and has been gone from different parts of North America something like 15,000 to 5,000 years. In terms of geologic time, present landforms are very young. Most of the erosional activities discussed in this book took place in the last few million years, many of them in the last few thousand.

CLIMATE AND GEOMORPHOLOGY

Climate controls, to a great extent, the types of geomorphic processes that work on the earth's surface and their rates of operation in a particular region. In the technical sense, *weather* involves the atmospheric conditions at the moment—wind direction, velocity, barometric pressure, clouds, precipitation, etc. *Climate* is the average in a particular spot or area of changing weather conditions throughout the seasons. Of the many parameters that might be used in describing climate, two measures are dominant: *mean annual temperature* and *mean annual rainfall* (or precipitation). The former is the yearly average of daily average temperatures, while the latter is an average of yearly rainfalls. Rainfall is the more significant item because it is the key factor in weathering, and runoff is the most important agent of erosion. Temperature is important because it largely determines the type of precipitation, whether freezing occurs, and, in part, it regulates runoff. Seasonal variations and fluctuations in temperature and rainfall produce a variety of climates.

Obvious examples of climatic control of geomorphic processes are found in glaciated and arid regions, where landscapes with distinct and unique characteristics have developed. Other climatic regimens, differing only slightly from the humid temperate, glacial, or arid climates, may also develop characteristic landforms and landscapes. Not enough study has been applied to this problem, but it is recognized that subtle variations that are probably climatically controlled can be found in landscapes over the world (Waipio, Hawaii, fig. 1.3).

MAGNITUDE AND FREQUENCY OF DYNAMIC FORCES

Is more geologic work (i.e., erosion, transportation, deposition, etc.) done during major catastrophic events, such as great floods, storms, etc., than is accomplished little by little through the continuous, day-by-day activity of geomorphic processes? Wolman and Miller concluded, in a discussion of this question (1960), that the major amount of sediment transport by streams in humid, temperate regions occurs during the high flows that come once or twice a year. Although extremely large floods carry greater quantities of sediment, these occur so rarely, that from the standpoint of transport, their overall effectiveness is less than that of the smaller, more frequent floods and seasonal high water stages. On the other hand, the progressive deposition by the overbank flows of these "normal" floods is not, apparently, responsible for formation and enlargement of a floodplain. Rather, the principal mechanism seems to be lateral movement of the channel and depositional activities that characterize the "routine" work of the stream.

Studies of the transportation of sediments by wind suggest that much of the effective work wind does on the landscape is performed during events of moderate magnitude and relatively frequent occurrence, such as dust storms over the plains in early spring. With regard to the force of waves, the observations of this writer and others who have studied modern beaches indicate that the significant changes in terms of erosion and deposition have usually been the result of one or two major storms a year. On Cape Cod (Massachusetts) beaches, for example, one severe northeast storm can move more sediments and do more damage than all the breakers of several summer seasons (Ziegler, Hayes, and Tuttle, 1959).

Figure 1.3. Waipio, Hawaii (1:62,500, C.I. 50 ft., 1916). Because of the heavy rainfall, many streams have developed on the windward side of the island of Hawaii. They flow down in nearly straight, almost parallel channels that radiate densely from the summit of the shield volcano Mauna Kea (Chapter 7). The steep valley side slopes are covered with lush semitropical vegetation, underlain by fairly uniform volcanic rocks. The parallel drainage pattern (Chapter 4) and the extremely steep slopes are the result of factors involving climate (primarily), vegetation, rock type, and geologic structure.

MAP INTERPRETATION AND TERRAIN ANALYSIS

Many of the earth's landforms are too large to be seen directly by the human eye, and the view from an aircraft is too fleeting for study. Thus maps, charts, aerial photos, photomaps of many types, and especially small-area, large-scale maps are the mainstays of students of land surfaces. Fairly accurate topographic maps, or *quadrangles,* with scales of one inch equaling a mile or larger, are available for about two-thirds of the United States, all of the British Isles, most of western Europe, and scattered spots elsewhere in the world. The portions of topographic maps included in this book are selected from quadrangle maps of *The Topographic Atlas of the United States,* published by the U.S. Geological Survey.

Map reading is an easily acquired skill by which one visualizes objects from symbols and perceives topographic and cultural features of importance to one's purpose. *Map study and interpretation* is a much more complex operation demanding special skills and qualifications. The human eye must be trained to see indications of lineation, types of patterns, lack of pattern, etc., in the appearance of landscapes, both in nature and as portrayed in various kinds of maps. From a study of the *terrain,* one can frequently recognize the nature of spatial distribution of features, although usually this needs to be confirmed by quantitative analysis because the eye may be deceived or may not be capable of recognizing the effect being looked for. Determinations can be made as to the probable origins of landforms, and hypotheses then developed as to the history and sequential development of landscapes. These are geomorphic approaches (Curran, Justus, Perdew, and Prothero, 1974; Upton, 1970).

Geologists depend a great deal on maps, particularly topographic maps, in making geologic interpretations from landforms and landscapes. Clues can often be found as to the nature of bedrock, its structure, geologic processes producing the topography, and the sequence of events making up the geologic history of a region. Some of the important uses for map and terrain studies are the following: route surveying (highways, railroads, pipelines, transmission lines, etc.); exploration and development of "mineral" resources (sand and gravel, rocks of economic value, petroleum, etc.); soil conservation and land management including forest reserves; water supply problems (flood control, dam sites, reservoirs, irrigation, pollution, sewage disposal, etc.); navigation (rivers, harbors, canals); urban planning, demographic studies, airport location, industrial site location, etc.; beach erosion and waterfront development; outdoor recreation and sports; lunar and planetary geology; terrain analysis for military purposes.

Remote Sensing

A cluster of new techniques in *remote sensing* have been added to the valuable procedures of photo interpretation since imagery can be received from satellites or aircraft. In addition to black-and-white and color photography, researchers can use side-looking radar, infra-red, airborne magnetometers (to measure variations in the earth's magnetism), and other instruments for special purposes. By use of remote sensing, flooded areas, crop damage, soil moisture, etc., can be quickly located and measured (Reeves, 1975). In *Terrain Analysis*, Way (1973) describes deductive geologic and geomorphic analogs and suggests how rural and urban planning can be adapted to existing landscapes and the natural environment, especially in terms of site selection, sewage disposal, bearing strength of regolith (p. 20), construction materials, runoff and infiltration, route location, and cropland utilization.

SUMMARY

Geomorphology: the study of landforms and landscapes, their origin, classification, and description.

Landform: a single terrain feature.

Landscape: an assemblage of landforms.

Geologic structure determines location and overall shape of landscapes.

Erosion $= \pm$ weathering $\pm$ mass wasting $\pm$ entrainment $\pm$ transportation $\pm$ deposition.

Most topography is post-Tertiary in age. Climate controls the type and rate of geomorphic processes. Yearly floods and storms, especially in humid, temperate regions, probably accomplish more erosion than an occasional great flood or storm.

Chapter 2

Weathering and
Mass Wasting

Dictionary definitions and some discussions of weathering tend to emphasize its destructive aspects and fail to point out that weathering is the geologic process most important to human existence. Simply to compare surface conditions on the moon with those on our planet (which has had the benefit of billions of years of weathering) gives an indication as to why this is so. Without weathering, we would have no soil for the nourishment of terrestrial life, very few nutrients in the sea for marine organisms, no sedimentary ore minerals (iron, aluminum, etc.), no fossil fuels, few, in fact, of the raw materials on which our civilization depends.

TYPES AND RATES OF WEATHERING

In a broad sense, weathering involves more than the *decomposition* (chemical decay) and the *disintegration* (physical breaking up) of rocks. Reiche (1950) has defined weathering as ". . . the response of materials which were in equilibrium within the lithosphere to conditions at or near its contact with the atmosphere, the hydrosphere, and, still more importantly, the biosphere. Its general nature is well indicated . . . as the change of rocks from the massive to the clastic state." Reiche's definition expresses in technical terms the concept of weathering as a group of processes interacting in the contact zone or interface of rocks, air, water, and organisms, with a variety of chemical and physical changes resulting. In a comprehensive sense, weathering prepares the rocks of the earth's crust for erosion.

Man has long known that some rocks show the effects of weathering sooner than others, and he has tried to choose for building purposes and for monuments those stones that could withstand exposure to the elements. However, a rock type that is considered "everlasting" in one climatic environment may crumble or disintegrate in another. Why are long-exposed outcrops of massive granite, for example, hard and resistant on Mt. Rushmore, South Dakota; broken and shattered on top of Pikes Peak in Colorado; and as soft as rotting cheese in the tropical climate of Hong Kong? Why does limestone in some places erode more rapidly than in other places—forming valleys in the Appalachian Mountain regions of Virginia, but boldly standing out as resistant rimrock on some mountains in Idaho and Utah?

In the main, five factors influence the type and rate of weathering processes: (1) climate, (2) rock and/or mineral composition and structure, (3) topographic position, (4) vegetation, and (5) time. In the case of granite, it would appear that warm surface water running down slopes in the humid, hot climate of Hong Kong is more destructive than the freezing and thawing attacking similar rock on top of Pikes Peak. Limestone, because of its chemical composition, decomposes more rapidly in a humid climate than in desert country. In both cases weathering tends to be more intensive on slopes than on level areas because the force of gravity helps to remove loosened pieces of rock, thus exposing a fresh surface to the weathering process. The amount and type of vegetation present, governed mainly by climate but to some extent by typography, may serve as a protective cover and thus reduce some kinds of weathering activity; but vegetation mainly provides an important, continuing source of decaying organic matter and acids that help along the chemical reactions of decomposition. The factor of time is significant when other conditions are stable in that more weathering is accomplished over longer periods of time.

In order to gain a clearer idea as to how weathering influences the development of landforms and landscapes, it is helpful to examine the various components to see how these processes work. At the same time, it should be kept in mind that in nature several types of weathering are usually working together on earth materials and, furthermore, continue to go on while sediments are being transported and deposited.

Effects of Precipitation and Temperature

Because weathering cannot take place in the absence of moisture, the availability of water in its various forms is a critical factor. Weathering processes may be speeded up or slowed down by changes in the supply of moisture, or by the rise and fall of temperatures, or by the two factors working in conjunction. The rate at which soluble materials go into solution is directly related to the amount of rainfall and the degree of humidity. Leaching (the moving downward of minerals in solution) is also intensified as temperatures rise because the rate of most chemical reactions increases in the presence of heat. A rise of only ten degrees Centigrade (18° F.) may increase the rate of chemical reaction two- or threefold. As this relationship applies also to many biological phenomena, the decomposition of organic matter in the soils of tropical regions, for example, proceeds at a faster rate than in the temperate zones (Thomas, 1974). Climatic parameters strongly influence both physical and chemical activity in a weathering zone (Ollier, 1969).

Porosity and Permeability

Porosity, the amount of open space in rock, soil, etc., is a limiting factor on the volume of water or air that can be contained in earth materials. This parameter is expressed in percent as the ratio of pore volume to the total volume of the material. Any porosity over 15 percent is considered high. The degree of porosity of rocks and other earth materials varies with bedding, jointing, sorting, packing, shape of grains, amount of cementation, etc. In general, clastic sedimentary rocks have more pore spaces and voids than do massive rocks unless the latter are fractured and jointed. Porosity is a significant factor in the storage of fluids in rock but is not *in itself* a critical aspect of weathering.

Permeability, the degree of ease with which water moves through the open spaces in rocks and earth materials, does not always correlate directly with porosity. Some shales have high porosity; but because of very small openings between grains, the permeability is low. Under normal pressures water does not pass freely through openings smaller than 0.05 mm.) In cold regions, fine-grained, moderately permeable rocks such as slates, schists, and some sandstones tend to be susceptible to frost action. Highly impermeable and extremely permeable rocks are more resistant to weathering by freezing and thawing because little water penetrates impermeable rocks; while in very permeable rocks, ice crystals can form, melt, and reform without creating any great stress.

The porosity and permeability of rocks and sediments largely govern the rate of movement and the volume of water that can pass through them. When both porosity and permeability are low (as in a dense-textured igneous rock with few joints and fractures), underground water might move less than a foot in a year. In highly permeable rocks or sediments, water might move several hundred feet a day. The weathering and erosive effects would be scarcely perceptible in the former case and fairly obvious in the latter (p. 132).

Effects of Size and Shape of Rocks

Weathering of all types usually goes faster when block or particle sizes are small or angular because more surfaces are provided for the chemical and mechanical processes to work on. Similarly, smooth or polished rocks, inasmuch as they provide less area at the surface, are more resistant than rough, pitted, or grooved rock. On intensely jointed or bedded rocks, weathering tends to operate faster than on massive rocks. Exposed corners of rock are more subject to weathering processes than are flat or sheltered surfaces because reactions can gain entry via the angular rock faces. Some rounded boulders have been shaped in this way, a process called *spheroidal weathering*. When weathering operates on smooth or flat surfaces, it may cause the rock to come apart in thin layers and spaul off in sheets. This activity is one type of *exfoliation*.

CHEMICAL WEATHERING

The usual division of weathering processes into two main categories —chemical and physical (mechanical)—is convenient for purposes of study and analysis. Field investigations, however, invariably reveal that chemical and physical processes and effects are inseparably blended, although, in general, the chemical phases are more important. Usually the physical and mechanical types of weathering are found to be assisting or reinforcing the chemical reactions. What follows is a brief discussion of the principal processes and effects of weathering, for the most part as these are related to landform development.

Chemical weathering comprises a large variety of reactions occurring among and between (1) the gases and fluids from the atmosphere, plus those of surface and subsurface water; and (2) the rocks, minerals, sediments, and organisms of the earth's crust. Hydrolysis, hydration, carbonation, oxidation, and simple solution are the means by which the decomposition of earth materials takes place. All the processes of decomposition have a sorting effect in that the more resistant or more stable compounds in a weathering environment tend to be retained as residues, while some clastic materials and soluble compounds are carried off.

Hydrolysis

In this type of weathering reaction the metallic ions are replaced by the hydroxyl radical. Hydrolysis (along with hydration) is the usual means by which clay minerals are formed. The following equation shows the reaction of the hydrogen ion (H) and the hydroxyl radical (OH) with orthoclase feldspar:

$$KAlSi_3O_8 + H^+ \quad + OH^- \quad \rightarrow \quad HAlSi_3O_8 + \quad KOH$$

Orthoclase + hydrogen ion + hydroxl ion yields "clay" + potassium hydroxide

The unstable aluminosilicic acid changes to colloidal substances that become clay minerals. The presence of acids in water—such as carbonic acid (common in rainwater), strong metallic acids (formed by oxidation of sulphide minerals), acids from clays, and plant acids provided by organic matter—tends to provide more and more hydroxyl radicals and ions, that, along with the colloidal clays, are powerful agents of weathering upon the unaltered minerals and rocks with which they come in contact.

Hydration

Adsorption of water takes place in hydration, as illustrated below in the equation for the conversion of anhydrite to gypsum:

$$CaSO_4 \quad + 2H_2O \quad \rightarrow \quad CaSo_4 \cdot 2H_2O$$

Anhydrite + water yields gypsum.

Hydration usually occurs along with other reactions, such as hydrolysis, oxidation, and carbonation. The simple addition of new compounds by hydration results in an increase in volume, which, in turn, brings about disintegration. This is an example of how chemical weathering sometimes produces a mechanical effect.

Carbonation

The chemical combination of carbon dioxide (CO_2) or the bicarbonate ion (HCO_3) with minerals or rocks is carbonation. Carbon dioxide in the atmosphere and in the soil dissolves or combines with water to form carbonic acid (H_2CO_3). Bedrock containing the elements potassium, sodium, or calcium (such as granite rocks) undergoes weathering by carbonation and hydrolysis, as shown below:

$$2KAlSi_3O_8 + 2H_2O + CO_2 \quad \rightarrow \quad H_4Al_2Si_2O_9 + \quad K_2CO_3 + 4SiO_2$$

Orthoclase + water + carbon yields "clay" + potassium + silica.
dioxide   carbonate

The potassium carbonate is dissolved in water and carried away in solution, leaving colloidal clay and silica.

Oxidation

When oxygen is added to or replaces other elements in mineral compounds, the reaction is called oxidation. A familiar example of this type of weathering can be observed after metal objects, left outside in the wet for a period of time, develop a rusty appearance. The effects of oxidation are noticeable in weathered rocks containing high proportions of the minerals amphibole, pyroxene, or olivine (all of which contain iron plus other elements in chemical combination with "silicates"). The oxidation of olivine illustrates this type of weathering reaction:

$$MgFeSiO_4 + 2H_2O \rightarrow Mg(OH)_2 + H_2SiO_3 + FeO$$

Olivine + water yields
magnesium hydroxide + silicic acid + ferrous oxide.

The combination of oxidation and hydrolysis has yielded magnesium hydroxide, silicic acid, and ferrous (iron) oxide from the olivine. By means of hydration, the ferrous oxide takes on water and more oxygen, forming the mineral limonite:

$$4FeO + 3H_2O + O_2 \rightarrow 2Fe_2O_3 \cdot 3H_2O$$

Ferrous oxide + water + oxygen yields limonite.

Oxidation reactions at the earth's surface usually involve organic substances (mainly from plant decay) as well as minerals and rocks. These weathering processes are significant in the development of soils.

Solution

Leaching, or solution, occurs on outcrops bathed in rainwater and in the action of subsurface water (ground water) percolating or washing down through the soil, extracting and removing soluble compounds produced by chemical weathering (fig. 7.7, p. 134). While some geologic materials, such as rock salt and rock gypsum, dissolve directly in water ("simple solution" reaction), materials in most weathering situations undergo combinations of hydrolysis, hydration, carbonation, etc., that produce ionized substances and contribute to greater chemical activity. The importance of the solution process is that it provides a mechanism for the release of ions and molecular units and also for the rapid removal of material. One of the most common solution reactions of chemical weathering is that by which calcium carbonate is dissolved, or leached, yielding soluble bicarbonate, as shown in the following equation:

$$CaCO_3 + H_2O + CO_2 \rightarrow Ca(HCO_3)_2$$

Calcium carbonate + water + carbon dioxide yields calcium bicarbonate.

The diagrammatic arrangement of minerals in Figure 2.1 is useful in evaluating the relative resistance or susceptibility to chemical weathering of some of the common minerals that are constituents of igneous and metamorphic rock. This series, based on observational data, illustrates the relative order of mineral stability, as proposed by Goldich (1938). The products of chemical weathering tend to be those that are most stable under the conditions of temperature and pressure at the earth's surface, the inference being that the response of materials to weathering tends to be in the direction of equilibrium.

The arrangement of minerals in Goldich's diagram is similar to that in Bowen's reaction series, which shows the order of crystallization of minerals in igneous rocks forming in a magma. (Olivine and calcium

Less resistant to weathering

OLIVINE

                                        Calcium plagioclase
    AUGITE (a pyroxene)
                                        Calcium-sodium plagioclase
        HORNBLENDE
        (an amphibole)
                                        Sodium plagioclase
            BIOTITE

        ORTHOCLASE
        (a potassium feldspar)

            MUSCOVITE

            QUARTZ

More resistant to weathering

Figure 2.1.   Mineral stability series.

plagioclase crystallize first at the higher temperatures; as the magma cools down, muscovite and quartz, which crystallize at lower temperatures, are the last minerals to come out). Keller (1957) has provided a systematic explanation for the similarity of Goldich's series to Bowen's. The crystalline structures of the first minerals to form in a magma are less complex and are not so tightly bonded as the later-cooling minerals. The energy necessary to bond together elements and radicals in minerals that make up igneous rocks increases from the beginning of the sequence to the end. Later on when an igneous rock is exposed to chemical weathering, these differences in the energies of formation of the constituent minerals apparently also control their thermodynamic susceptibility to decomposition. For example, olivine, having a lower energy of formation, tends to be more susceptible to chemical weathering than augite, and so on down the series to quartz, which is chemically inert. Rocks are compared by evaluating the resistance of their mineral components. A gabbro, for example, composed of plagioclase, olivine, and hornblende, can be expected to weather chemically faster than granite (orthoclase and quartz). A mica schist, composed mainly of muscovite and biotite (plus other minerals), weathers chemically faster than a quartzite, which is essentially all quartz.

PHYSICAL (MECHANICAL) WEATHERING

The processes of physical weathering render rocks more susceptible to chemical attack by breaking up and thus increasing the surface and interface areas. Much of this activity takes place concurrently; i.e., rocks disintegrate while they are decomposing. When chemical weathering increases the bulk of a rock substance, for example, the resulting strains and stresses tend to loosen up more cracks, joints, etc., which become pathways for further chemical activity. Other types of physical weathering include (1) rock expansion due to unloading, (2) frost action and other types of crystal growth, (3) organic activity, (4) colloidal plucking, and (5) thermal expansion and contraction.

Unloading

When massive rocks are uncovered by the removal of several thousand or more feet of sediments and rock, the relief from this confining pressure causes expansion and leads to the development of sheet joints or fractures, roughly parallel to ground surface. The joints may be less than an inch apart at the surface but many feet apart when deep in the bedrock. The effect of the expansion on surface outcrops causes them to exfoliate or peel off in concentric sheets. This phenomenon has occurred on a large scale in some areas, as in Yosemite National Park, where huge exfoliation domes rise above Yosemite Valley.

Frost-Wedging or Frost-Heaving

Another type of expansion causing rock disintegration is brought about by crystallization, usually from water freezing into ice but sometimes from salts coming out of solution in arid climates. The action of ice crystals is more significant, of course. The maximum effectiveness of freezing water occurs when ice is completely confined, causing tremendous pressures to be built up. In the usual type of situation, physical weathering is most extensive when alternate freezing and thawing occurs so that ice melts and refreezes along fractures and joint planes. As might be expected, where plenty of water is present, frost action on exposed rock is severe. Even under the most intensive conditions, however, frost action is largely confined to areas of rock at or near the surface. Moderately high mountains in areas of average rainfall may be subjected to frost action so intense that the upper slopes become completely buried under blankets of broken and shattered rocks. Such an accumulation of mechanically weathered blocks of rock is called a *felsenmeer* (from the German "rock-sea").

Organic Activities

Although organic weathering is largely chemical, physical components are usually involved. Root pry and tree growth may widen cracks in rock. Burrowing animals and earthworms dig up and rearrange loose particles of rock and soil, thus exposing fresh material to weathering by air and water.

Thermal Expansion and Contraction of Rock

Because rock is a poor conductor of heat, the exposed outer parts of boulders or outcrops can become very hot in the direct sun without much temperature change occurring a few inches beneath the surface. In climates where the daily range is from 30 to 90 degrees Fahrenheit, continual expansion and contraction caused by heating and cooling of outer parts may loosen the outer layer of a rock. Moreover, in coarse-textured rocks that contain several different minerals, these components may expand and contract differentially (different minerals have different coefficients of expansion), causing individual grains to loosen. This thesis seems logical, but laboratory tests involving many cycles of heating and cooling have failed to bring about physical disintegration of rocks. Although it is true that rocks sometimes break in the heat of forest fires and campfires, field studies in the hotter areas of western North America have failed to turn up evidence that thermal expansion and contraction *by itself* causes rocks to disintegrate. What may happen is that stresses set up by expansion may produce minute cracks, into which air and water vapor can penetrate, thus initiating chemical weathering.

Colloidal Plucking

The importance of colloidal plucking as a weathering process has not yet been determined. It is believed that the contraction of soil colloids when drying may exert a strong pull which might flake off or loosen mineral grains or very small pieces of rock. Plucking may occur after each wetting and drying of soil colloids.

ENVIRONMENTAL EQUILIBRIUM

Most of the rocks of the earth's crust were formed under conditions of considerably higher temperatures and pressures than those prevailing in the interface zone where weathering takes place. The changes that occur during weathering proceed toward stability of rocks and minerals in environments of low temperature and pressure. (The opposite situation, the change of rocks and minerals formed in surface environments into new compounds in environments of higher pressures and temperatures, is called metamorphism.) According to Le Chatelier's principle, if a stress—such as a change in concentration, pressure, or temperature—is applied to a system in equilibrium, the equilibrium shifts in a way that tends to offset the effects of the stress. The dynamics of weathering follow this principle. The lower pressure at the earth's surface encourages the formation of substances of greater volume and lower density (i.e., mineral hydrates, oxides, carbonates, etc.) than the original pyromorphic (heat-formed) minerals being weathered. Also, in accordance with Le Chatelier's principle, the chemical reactions of weathering are exothermic, or heat-liberating. In a global sense weathering can be visualized as a complex means by which materials adjust to an environment.

EXAMPLES OF HOW WEATHERING WORKS

The weathering of granite illustrates how processes of decomposition and disintegration operate. Granite bedrock usually has several systems of intersecting joints or cracks formed in the geologic past when the rock was subjected to stresses deep in the earth's crust. Later on, after many thousands of years of erosion, sheet jointing developed due to expansion following unloading. The seemingly strong and solid granite bedrock had within its mass numerous cracks and incipient planes of weakness. Into these cracks, water and air penetrated, causing further opening and widening of the joints through a combination of chemical and mechanical processes. With time and the aid of gravitational force, parts of the rock separated along the joint planes, forming distinct blocks or *boulders* that, in turn, split apart and were rounded off by long continued weathering.

Mineralogically, granites are composed of orthoclase feldspar and quartz, with or without minor amounts of muscovite, biotite, plagioclase, augite, and hornblende. Expansion due to chemical and physical processes causes the mineral grains to become slightly loosened and separated from each other. Some outcrops of granite are surrounded by blankets of loose fragmental rock and mineral grains called *grus*, that are produced by weathering processes.

The quartz grains do not change chemically but eventually become *sand*. Orthoclase feldspar, being only moderately resistant, undergoes hydration, hydrolysis, and carbonation, decomposing into a *clay* that is insoluble, a soluble carbonate, and a soluble silica. The *soluble compounds* are carried away in solution and may be later deposited as intergranular cement, as cave or spring deposits (travertine or sinter), or as chemical sediments or oozes in the oceans (where some may eventually be converted into limestones or cherts). Thus the weathering of granite under differing environmental conditions may produce nearly all sizes and types of sediments: boulders, grus, sand grains, clay, and soluble compounds. The relative proportions of types of sediments depend on the composition and location of the bedrock and the length of time it has been exposed to weathering, as well as the climatic environment.

The weathering of other kinds of igneous rocks produces different amounts of sand and clay, depending on the original mineralogical composition. Rocks high in feldspar yield more clays. The mafic rocks, which contain no quartz, do not yield quartz sand but may produce material of sand size. The decay of carbonate rocks yields, in addition to soluble compounds, a small amount of insoluble silica (chert and sand grains) and insoluble iron compounds. Sandstones and shales weather to form sand, silt, and clay.

Through geologic time, rocks that have been broken down by weathering have provided material for the formation of new rocks, which, in turn, are subjected to weathering as they become exposed The sediments and soils of the earth's surface can be thought of as somewhere in the middle of this endless cycle.

Landforms Produced by Weathering

In a technical sense, it may be inaccurate to refer to any landforms as having been produced by weathering because generally they are the result of weathering in conjunction with some agent or agents of erosion. Certain topographic features are, however, mainly the result of differential weathering that has caused resistant beds or structures to stand out in relief. Conversely, where weaknesses exist, weathering attacks with great effect, producing holes, trenches, etc. The appearance of features resulting from differential weathering is apt to be so striking that many have been given fanciful or colorful names, such as Devil's Slide near Interstate 80 in Utah.

REGOLITH AND SOIL

Regolith is a term used to mean any weathered or unconsolidated material overlying bedrock. *Soil,* a type of regolith material that is characterized by layers (horizons) parallel to the surface, has been so modified by physical, chemical, and biological processes that it can support plant growth. (In engineering usage the term soil is used as the equivalent of regolith to mean any loose surface material, such as that which might be encountered in highway or foundation work.)

A *residual* regolith grades downward through weathered rock to unweathered, solid, parent bedrock of similar mineral and chemical composition. *Transported* regolith usually has a sharp contact with underlying bedrock (which may or may not be weathered) and contains, in most instances, different minerals. Bedding, sorting, or other evidence may indicate that transported regolith has been moved from its original source. Regolith comprises—besides soils—many types of rock wastes and organic accumulations, all being acted upon by weathering environments.

The formation of soils, however, involves more than the weathering of rock. Determining factors in the development of soil types are (1) climatic environment, (2) organic components, (3) topography, including hydrologic features, (4) parent material, and (5) length of time of soil formation. These factors are so interdependent that change in any one affects all the others. It is interesting to note that in some respects the scientific view of soil as a dynamic system, constantly evolving, is not unlike ancient beliefs, based on naturalistic observations, that ascribed to soil a life and rhythm of its own.

The study of soil and soil-forming processes, aside from obvious economic significance, has contributed valuable information for an understanding of landscape development because soils reflect the geomorphic history of an area. To a geomorphologist, soil maps and profiles, along with topographic maps and aerial photographs, are major tools in the study of landforms. The type, extent, and characteristics of a soil or soil series may be helpful in determining the following: Type and/or source of the parent material, climate existing during soil formation, topography of the surface during soil formation, length of time for development of the soil, type of vegetation present while soil was being formed, degree of erosional activity, and possible identification from composite soil profiles of changes in climate, vegetative cover, etc.

MASS WASTING OF SOIL AND ROCK

Gravity provides the dynamic force for interacting geomorphic processes that bring about *mass wasting*, which is the term for slow or rapid downslope movement of regolith and bedrock. On valley sidewalls, rocks are weathered and broken up; materials are loosened and steadily delivered to a stream which carries the debris away. Although, strictly speaking, stream erosion is limited to the actual stream channel below water level, the removal of these products of weathering and mass wasting is an important function of rivers and streams. Within the expanse of geologic time, processes of mass wasting may cause an individual particle to move once every few centuries, once a year, or every day. At the other extreme, some rockfalls may have velocities of over 100 miles per hour.

Creep

A slow downhill movement occurs even in regions where trees and vegetation bind soil and rock debris together (fig. 2.2). Evidence of *creep* can be seen on hillsides where tree trunks, fenceposts, telephone poles, or old gravestones are tilted downslope. In some places, stratified bedrock has been bent and moved in a downslope direction.

Earthflows, Mudflows, and Solifluction

A slightly more rapid type of movement tends to occur on steep slopes in regions of humid climate. Heavy rains or melting snow soaking into the ground increase the weight of the regolith, causing hillside material to flow downslope as a muddy, tumbling mass of wet soil, boulders, trees, etc., moving like a sluggish stream (fig. 2.2). Large flows can be destructive, especially when they come down from steep canyons onto a plain. *Solifluction* is a slow type of earthflow involving the movement from higher to lower ground of masses of soil and waste saturated with water from rain or melting snow and ice. Flow may occur on slopes as flat as four or five degrees, with velocities up to three inches a year. Where optimum conditions prevail year after year, slopes are reduced considerably, even at this slow rate.

A more intense type of solifluction operates in periglacial regions where the ground freezes either seasonally or permanently. When thawing occurs in the upper levels of the ground, the melting soil and weath-

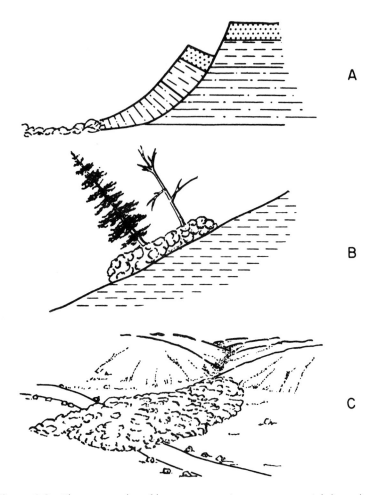

Figure 2.2. Three examples of how mass wasting moves material downslope:
A. Slump. The material in the moving block is rotating slowly down and outward in relation to the stationary bedrock. At the bottom of the slide, material is pushed out to form a hummocky toe; at the top, a step or small
terrace is evident. **B. Creep.** Here loosened regolith edges slowly downward
over stable material below. **C. Mudflow.** Moving material is churned and
mixed as it flows at a perceptible rate out of a small valley. (From **Physiography of the United States** by Charles B. Hunt. W. H. Freeman and Company.
Copyright © 1967).

ered debris become mobile because the water cannot percolate down below the thaw area. The wet top material shears off and slides downhill over the still frozen, impermeable ground beneath. As fresh material is exposed, another layer soon becomes mobile under the hot sun and rains typical of "thaw seasons." Detailed studies by Washburn (1965, 1967) of mass wasting on the east coast of Greenland indicated that both solifluction and frost creep were operating on bare mountainsides to produce a slow, ratchet-like downslope movement. There the creep was due to seasonal rather than diurnal variation of temperature, and moisture was the controlling factor for solifluction, which did not proceed until the regolith was saturated. Soil movement varied according to type of material, with coarse-grained sediments having high porosity showing little movement. Finer-grained sediments that tended to confine or retain water were more susceptible to frost creep and solifluction. These processes of nearly imperceptible flow tend to produce extensive terraces and lobes having a stepped appearance (see p. 112).

Slump

Slump is a slow to moderately rapid type of mass wasting by which a section of regolith or bedrock is moved downslope in the form of a large mass or block that slides downward and outward along curved fracture planes, causing the block to rotate slightly and producing a scarp along the back of the block with subsidence on the top. Slumping is usually caused by oversteepening of the base of a slope—as by wave erosion, bank-cutting by streams, building construction, road cuts, etc. If the earth material does not have strength enough to hold its original position; under the pull of gravity, the blocks slide and rotate downward along fracture planes (fig. 2.2).

Landslides

Avalanches, slides, and falls are spectacular types of mass wasting characterized by sudden rapid movement. As with slump, landslides tend to be the result of undercutting. An unsupported mass of rock and debris may slide, fall, or bounce down the slope, frequently at high speeds, overriding or burying everything in its path. If conditions are such that landslides may occur, preventive measures can be taken. Because water is usually the principal cause of avalanching, an incipient landslide mass should be kept dry and stable. This is usually accomplished by a diversion of drainage above the potential slide area and by putting drainage pipes within the landslide mass to carry off ground water, thus stabilizing the toe of the slide.

On relatively steep slopes where bedrock outcrops, weathering, especially frost heave, loosens blocks and chips of rock. Under the influence of gravity, these chunks and pieces accumulate at the foot of the slope as *talus,* or scree, in the form of thick sheets or cones. Talus usually stands at the maximum angle of repose for large-sized, angular material and forms significant landforms in mountainous topography. Where talus has become saturated with water, snow, or ice, it may move slowly downward in a stream-like manner. Such moving masses of rock are called *rock streams,* or *rock glaciers.*

Air-Layer Lubrication

A recent examination of landslide breccia at Blackhawk Mountain, California, has produced a new theory about a possible mechanism by which large amounts of material can be moved swiftly considerable distances downslope. In prehistoric time the Blackhawk landslide began as huge rockfalls, and the falling rock was launched into the air after passing over an obstructing ridge. Below the ridge the airborne rock mass traversed relatively gentle, smooth slopes without materially disturbing the underlying rock. It is postulated that the undeformed sheets of limestone breccia moved on a layer of air trapped under the slide rock; i.e., by air-layer lubrication. Speed of the slide may have been in excess of 100 miles an hour. Comparisons with similar landslides, such as rockfalls, suggest that both the Elm rockfall, which occurred in Switzerland in 1881, and the Frank, Alberta, landslide of 1903 moved by air-layer lubrication (Shreve, 1968).

Landslide Topography

A variety of landforms and a characteristic landscape—landslide topography—are produced by mass wasting. Creep tends to produce lobate bulges on hillsides. Solifluction builds up accumulations of material at the base of slopes and on valley floors, that later may be cut into terraces. Slump leaves scarps along the backs of blocks; and if multiple blocks slide down, they may form a jumble of features, with patches of slope facing different directions. Where slumping is active in unconsolidated materials, downslope movement brings about the formation of the so-called "cattle steps," or "animal tracks," often seen on hillsides. Animals (and humans) may use these tracks, but they are usually the result of slump rather than biological activity. Sometimes small lakes form at the back of slump blocks, for the surfaces of slump block areas seldom have organized patterns of drainage.

Flows of mud or debris often form lobate ridges, sometimes with distributaries, as a mass of debris stops moving and begins to dry out. Sometimes the edges of flow stop moving while the center continues to move, creating parallel ridges that resemble natural levees. Scars bare of soil and vegetation are left after a flow has swept down a hillside or gully. Landslides leave large scars or great, bowl-shaped depressions on mountainsides or hillsides. Below, where large landslides have come to rest, the typography is a jumble of hillocks and depressions without organized slopes or drainage (fig. 2.2).

SUMMARY

Weathering prepares bedrock for erosion and plays a key role in the breaking up, modifying, and removal of earth materials, as well as the formation of regolith, soil, and a variety of landforms. Five factors determine types and rates of weathering: (1) climate, (2) rock and mineral composition, (3) topographic position, (4) vegetation, and (5) time. In any region, answers as to why a particular rock is resistant or non-resistant to weathering can be sought in the "mix" of factors controlling the weathering processes to which the rock is exposed. Types of chemical weathering include hydrolysis, hydration, carbonation, oxidation, and simple solution; physical (mechanical) weathering is mainly the result of expansion due to unloading, expansion from chemical effects, frost action, organic activity, colloidal plucking, and thermal expansion and contraction. Mass wasting contributes to the reduction of uplands by moving materials downslope.

Landforms Resulting from Weathering

Exfoliation domes, talus, felsenmeers.

Landforms Produced by Mass Wasting

Talus slopes, talus cones; rock streams; earthflows, mudflows; avalanches; landslide scars, landslide deposits. (Erosional scars are left on hillsides; piles of material form deposits below.)

Example of landslide topography: Hegben Lake earthquake and landslide area (Montana, west of Yellowstone Park).

Chapter 3

Stream Growth and Valley Development

TOPICS

The hydrologic cycle
Hydraulics
How streams erode and transport material
Factors controlling the rate of stream erosion
Base level and the graded stream
Valley shapes and the retreat of slopes
Alluviating streams and floodplains
Meandering rivers
Deltas, fans, and terraces

Scarcely a landscape exists that does not have some features, either erosional or depositional, produced by the action of streams. Furthermore, an understanding of stream flow (hydraulics) can be exceedingly useful in the study of glacial ice, the wind, and ocean waves because, in a fundamental sense, all behave as fluids and react according to basic physical laws of fluid mechanics.

Stream, the more or less universally used word in English meaning any watercourse or river, has many locally favored equivalents. In New England, people talk about *brooks.* In New York and New Jersey, the Dutch word *kill,* meaning stream, is found in a number of place names, such as Katerskill and Arthur Kill (name of the river behind Staten Island). Bull Run, the site of two famous battles in the Civil War, is a stream in Virginia, the term *run* being common in the middle Atlantic states. Farther south, small streams are called *branches,* as indicated in the classic phrase, "bourbon and branch [water]." The Cajun French word *bayou* refers to small waterways along the Gulf Coast. Dry, or intermittently dry, stream valleys in the arid Southwest are called *arroyos* or *washes.* Similar features in eastern Washington are called *coulees.* Throughout most of central and western America, the word *creek* is used to mean a small stream. However, pronounce creek to rhyme with "peek" in most of the Midwest, but remember to say "crick" in the western mountains. (Several of the synonyms for stream given here can be found on the map illustrations in this book.)

THE HYDROLOGIC CYCLE

In a planetary sense, there is only so much water on, in, or around the earth, much of it always in motion in one or more physical states as gas, liquid, or solid. Constantly working on the water are two main driving forces: (1) *solar energy*, which causes evaporation from land and sea surfaces, and through the action of winds, moves water vapor great distances, and (2) the attraction of *gravity*, which causes water to flow downhill. Minor additional movements are the result of capillary action, hydrostatic pressure, plant metabolism, and volcanism.

Because of the repeated changes in state and location, movement of the earth's water is described in terms of a hydrologic cycle involving the following phases (not necessarily in this order):

Evaporation of sea water into the air.

Movement of water vapor in the atmosphere.

Precipitation (rain, snow, sleet, etc.).

Evaporation from land and inland water surfaces.

Runoff of water over the land.

Infiltration of water into the crust of the earth and subsequent percolation.

Discharge of ground water into streams, lakes, etc.

Transpiration by plants of water from the earth into the atmosphere.

Accumulation and storage of water as snow and ice in glaciers, followed by melting.

As long as energy-laden sunlight continues to arrive on the planet earth, the hydrologic cycle will provide running water, which, as a continuous agent of erosion, produces change in the interface zone of land, oceans, and air.

Ground Water and the Water Table

The catchment and storage functions of *ground water* (subsurface water) make it a vital natural resource. Ground water supplies are "recharged" (renewed) as water from rain, sheetflow, streams, ponds, etc., sinks in and seeps down into the earth. When the available pore space, decreasing with depth, becomes filled, the rocks and regolith are said to be saturated. The irregular upper surface of this saturated zone bulging upward under hills and flattening under valleys is the *water table*. Above it, extending to the surface, is the zone of aeration where intensive weathering takes place (p. 15). This zone, which is unsaturated, contains air spaces through which ground water, under the influence of gravity, moves downward. Once water passes below the water table, it may move up, down, or laterally under the effects of hydrostatic pressure or gravitational force. During times of drought or low stream flow, ground water seeps into streams, adding to their volume; and when streams are in flood, some of the excess water goes back into the regolith and bedrock. This process of discharge and recharge tends to moderate the amount of water in streams, especially in times of low rainfall, when water enters streams though the bank channels.

HYDRAULICS

In understanding how a moving fluid (water) erodes and transports material, a few generalizations from hydraulics (the study of moving fluids) are helpful. Stream flow is, in most cases, what would be called *turbulent*. Very slow, undisturbed flow is described as *laminar*. In a typical turbulent flow, the individual paths of water particles intermix in a chaotic movement of eddies and flow patterns. Extremely turbulent flow, such as that found in rapids and waterfalls, exhibits swift shooting movements and is characterized by rapid changes in velocity over short distances.

The maximum velocity of water flowing in a channel is usually found near the center and only just under the surface. Areas of maximum turbulence are near the bottom yet away from the channel walls and floor because friction checks velocity and turbulence at the sediment-water interface. If a channel is irregular or obstructed, the areas of maximum velocity and turbulence shift around. If the cross-section of a channel is asymmetrical, the maximum velocity shifts toward the deep side and the zones of high turbulence may be displaced toward the sides of the channel (Morisawa, 1968; Schumm, 1972).

HOW STREAMS ERODE

Running water in a stream brings about entrainment of material in several ways: by *hydraulic action,* by *abrasion* (from particles already in motion), by *solution,* by the *vortex action* of eddies, and, in cases of very high velocities, by *cavitation.* Hydraulic action (or "hydraulicking") is the direct effect of particles being swept along by moving water, as happens when one hoses down a driveway with a swift stream of water. Cavitation is a process by which tiny bubbles of water vapor form in stretches of swift-flowing water and then collapse when the channel widens out and the velocity decreases. The collapse of the bubbles causes shock waves to be sent through the water, hitting the banks and bed of the stream with hammer-like blows. Cavitation is relatively rare in natural streams because it occurs only where the water flows at a very high velocity.

Whether or not a particular grain or rock fragment gets entrained (picked up) from the bed of the stream depends, first of all, on its size. If a grain is too small, it will not stick up enough into the moving fluid to be in a zone where velocity is sufficient to start it moving. Rounder particles and those with lower specific gravities are moved more easily along the bed of the stream than flat or angular particles or those with high specific gravities. Location is a factor because in the so-called boundary conditions at the bottom of a stream, more velocity is needed to entrain a particle than to keep one moving after it has been swept up into faster flowing water. Once suspended in the stream flow, angular lighter particles can move faster and farther; when in motion, they can also start other particles moving by hitting them.

HOW STREAMS TRANSPORT MATERIAL

Streams transport sediments by rolling them along the bed (*traction*), by bouncing particles and causing others to jump when hit by a moving particle (*saltation*), by means of mechanical *suspension,* and through chemical *solution.* The first three methods are dependent on the velocity of the water; that is, the faster the water moves, the more material and the larger the particles that are swept along. *Bed load* is the term for the solids carried by mechanical means, while the materials in solution make up the *dissolved load.*

Entrainment and deposition of the bed load are going on continuously as the speed of the water increases and decreases. So sensitive is this relationship that very slight changes in velocity have significant effects on the erosive power and transporting ability of a stream. The size (expressed in terms either of weight or volume) of material that can be moved may increase (or decrease) up to a maximum of the 6th power of the increase (or decrease) of the velocity of the stream. If a stream flowing two miles per hour increases its velocity to four miles per hour, the weight of small pebbles being transported (perhaps by traction and saltation) might increase from 0.2 grams up to a maximum of 6.4 grams. Extensive study of this relationship (called the *6th power law*) indicates that it does not hold completely for particles of fine sand or silt smaller than about 0.25 mm. because of boundary layer conditions (Rubey, 1938). In quantitative terms, the approximate minimum velocities necessary for the movement of the size particles indicated are (Sundborg, 1956):

Particle size (diameter)	Minimum velocity for beginning entrainment
0.001 mm.	300 cm./sec.
0.01	80
0.1	30
1.0	40
10.0	175
100.0	400

From these data, it can be seen that sand (0.1 to 2.0 mm.) is eroded relatively easily, whereas the finer clay and silt and the coarser gravel need a higher velocity for entrainment. Once particles of silt and clay (which are hard to start moving because of their small size) are entrained, they can be carried with less velocity, either in suspension or by saltation, than sand-sized particles.

The interrelationship between entrainment, transportation, and sedimentation is also influenced by varying shapes of particles, their specific gravities, and the position of particles on the bed of the stream. When many different sizes and shapes occur together, they may be easier to move than if they are all uniform in size. A few large grains cause local turbulence and an increase in velocity, hence an increase in erosion. Also, the presence of small particles on the bed of the stream may make it easier for larger pebbles to roll along over them.

FACTORS CONTROLLING THE RATE OF STREAM EROSION

With velocity as the key factor in the entrainment of material and its continued transportation in a stream, it becomes evident that a very slight change in velocity affects a stream's ability to erode and transport. Factors causing change in velocity are *volume, gradient, shape of channel cross-section, roughness of the bed,* and the *quantity of the load.* These variables are interrelated and affect each other in ways that may increase or decrease velocity, as shown in the generalizations below:

An increase in volume increases velocity, thereby causing more erosion.

An increase in gradient (whether by artificial or geologic means) causes an increase in velocity. Gradient (customarily defined as the average vertical drop per horizontal mile) is a numerical expression of the downstream slope of the water surface of a stream.

If the cross-sectional area of the channel is increased, velocity decreases, resulting in deposition.

An increase in roughness of the walls (banks) and bottom of the channel decreases velocity.

An increase in the amount of load decreases the stream's ability to erode and transport because energy used in carrying the greater load is no longer available for erosive action.

Man's activities often alter a stream's power to erode. A bridge opening that is too narrow increases velocity, and then the force of the water may wash out bridge piers. On the other hand, artificial widening of a channel may result in considerable deposition. The construction of a dam on a river creates a reservoir of slow-moving water, in which the stream deposits large amounts of its load. Downstream from the dam, the river has little if any load at first and may, therefore, begin to erode its channel.

STREAM REGIMENS AND FLOODS

Because flow characteristics of streams, especially the volume or discharge, vary considerably with the seasons and other changing conditions over a period of time, the collection of data about these changes becomes very important. In this country the U.S. Geological Survey builds and maintains gaging stations (small cement structures built on stream banks) that house automatic recording apparatus. Scattered at strategic points throughout a drainage network, these stations provide continuous records of water depths. From time to time the velocity also can be measured at the gaging sites. From these data, discharge, as it varies with time, can be calculated and expressed as a hydrograph, or rating curve, for a given stream at a particular gaging station. Analysis of rating curves produces a series of flood values of decreasing frequencies; i.e., yearly flood, average flood, 5-year flood, 10-year flood, etc. Some theoretical studies have used a calculated 1,000 or even 10,000 year flood.

A common stream regimen is one with low flow during the winter, followed by a flood peak in the spring, then tapering off, with low flow again in summer. During dry seasons, water in some stream channels comes largely from ground water discharging through the banks (p. 28).

A natural conclusion is that stream regimens differ markedly in different climates. Some rivers have winter floods, some have spring floods (perhaps from melting snow), and others have summer floods. In rare cases rivers have two separate flood seasons. Some streams in warm humid areas have very few floods, and the ones they do have are related to great storms, such as hurricanes or typhoons. Monsoonal climates produce distinctive patterns of flow because many streams receive essentially no runoff for periods of several months. In moderate to cold climates, spring runoff may be usual or abnormally high, depending on whether the ground stays frozen until late in the spring or thaws out early. Warm spring rain, falling on unmelted snow over frozen ground, produces heavy runoff and sometimes high floods. Heavy rains falling on unsaturated rocks with high infiltration capacities and a low water table may result in almost no flooding. In addition to factors of climate and rock permeability, the size and collecting capacity of the watershed area influences the amount of flooding. In the case of the lower Mississippi River valley in south central United States, rainfall and runoff in the upper Missouri River valley may either cancel out or reinforce flood patterns coming downstream from the upper Mississippi or Ohio River valleys.

Among the world's rivers, endless possibilities of flow variation may occur; but wherever a stream is located, sometime it is going to flood. Discharge may increase only a small amount, or may increase tenfold. Small streams, to be sure, show a proportionately greater increase in discharge during floods than do large rivers. What is certain is that at such times, great amounts of erosion, transportation, and (in the declining phases) deposition will occur. In narrow-bottomed valleys, the increased geologic activity brought on by floods may be intense, but is confined to a relatively small area. In broad valleys, flood waters may spread over miles of valley floor, occupying former channels and distributing great volumes of sediments. With the size of the channel greatly expanded during times of flood, the cross-sectional area is also increased many times. The augmented size plus the friction of obstacles, both natural and artificial, cause the flood waters to slow down, once they are outside the deep channels. Deposition then takes place, blanketing the floodplain as the velocity of the water decreases.

Ancient man welcomed the seasonal high waters of the Nile, the Tigris, and the Euphrates because overbank floods spread fresh sediments full of organic and "mineral" nutrients over the fields, thus renewing their fertility. On the floodplains of the present day, however, most of which tend to be densely populated, the disruptive effects of major floods are tremendous. Besides driving people from their homes, flood water buries gardens and fields, erodes new channels, tears out bridges and roads, undermines or washes out buildings and causes loss of life, both animal and human, as well as enormous loss of property.

Accounts of misery, famine, and death brought about by disastrous floods on the Hwang Ho (Yellow) River of North China are almost beyond belief. For many centuries the governments of China tried to control these floods; hordes of coolies, with barrows and baskets, built up great levee systems intended to hold the river within its main channel. When, however, an alluviating river is kept within its channel, the vast load of sediments, which normally would have been spread out as alluvium over the floodplain, are deposited on the stream bed, thus decreasing the cross-sectional area of the channel. This makes it easier for a subsequent flood to overtop the levees (fig. 3.2). Man's reaction to this problem was to mobilize greater numbers of people to build the levees higher until the river flowed "up in the air" at elevations above the floodplain surface, oftentimes at tree-top level. Then when a major flood occurred and the high levees became undermined or overtopped, the entire volume of the river poured out over the floodplain, inundating thousands of square miles and devastating the countryside. By the time the flood had subsided, the river had cut a new channel somewhere else on the floodplain, leaving the former channel high and dry.

Modern engineering practice tends to follow the pragmatic philosophy of "If you can't fight 'em, join 'em." Through the use of empirical and theoretical equations developed by hydrologists, the carrying capacities of all channels of a river system are calculated. Rainfall data and hydrographs are continuously monitored so that seasonal patterns of precipitation and discharge can be charted for each river basin. With the passage of years, integration and interpretation of the collected data enable engineers to predict with increasing accuracy the height or "crest" of an expected flood and its time of arrival at any part of the basin.

Once the maximum capacities of the river channels have been calculated, when more water than they can carry is expected to come down, floodgates in the levees are opened so that land of low value is flooded or auxiliary channels are filled, thus preventing the main river from overtopping the levees that protect cities, power plants, railroad yards, and vital installations. The strategy of evaluating the flood potential and yielding some areas while protecting others can be carried out only with the aid of efficient data-gathering services and a central controlling authority which can and must make decisions to flood some areas in order to save other more valuable ones. In the United States, flood control activities are carried out cooperatively by the U.S. Geologic Survey, the U.S. Weather Bureau, the U.S. Army Corps of Engineers, and agencies of the states involved.

Long-range preventive measures to eliminate the ravages of floods utilize the principle of enabling more water to infiltrate the earth's crust, rather than running off, especially during times of snow melting or heavy rainfall. Watersheds are reforested, gullies and uncultivated areas are planted with grass and shrubs, and artificial ponds are constructed on small tributaries. In the mid-length areas of stream systems, flood control reservoirs are designed so that they can be almost emptied prior to times of high runoff and then allowed to fill during the flood season. Water thus collected and stored is released gradually under controlled conditions during times of low runoff. Sometimes the operation of these reservoirs for the purpose of flood control conflicts with other human desires and needs for activities such as recreation, generation of electric power at the dam sites, early planting of crops, maintenance of navigation, and so on. To properly administer such multipurpose dams amid conflicting pressures from numerous groups requires competent analysis and the firm making of decisions based on the principle that allowing some damage and inconvenience to a few may prevent more serious damage or loss to many.

STREAM VALLEYS

Stream valleys are the most common and most important landforms on the earth's surface. Like the streams that shape them, valleys have many names—some common (gully, ravine, hollow, draw, bottom), some picturesque (gulch, flume, canyon, couloir, wash), and some poetic (vale, dale, glen, dell). All were cut by running water and are—or were—occupied by streams. Some troughs and valley-like features (*e.g.*, Death Valley) contain streams but were not formed by them. Such features bear the designation "valley" because of long-accustomed usage but have quite a different geomorphic history than stream valleys.

Base Level

In order to understand the overall development of stream valleys and stream-sculptured landscapes, we need an answer to the question: How far down can a stream (or a stream system) erode? The answer is—to *base level*. This is the limiting surface below which streams cannot erode, ultimate base level being closely related to sea level. Because water cannot flow across a horizontal surface, base level cannot be truly level. Its slope can be imagined as somewhat less than the gradient of either the lower Mississippi River (about one foot per mile) or the lower Amazon, approximately three inches per mile.

In using the concept of base level to visualize the amount of material that might be removed by erosion, one starts with the elevation of sea level and draws an imaginary, slightly sloping extension of this surface inland. This is a theoretical limit of erosion because it is unlikely that the crust of the earth would stand still long enough for erosion to go that far, nor would sea level remain the same over long periods of time. Valley widening and the headward erosion of streams are limited, not by base level, but by the development of divides and interfluves. These relationships, which have to do with the development of drainage systems, are discussed in Chapter 4.

The Graded Stream

Like the idea of base level, the concept of the *graded stream* is a theoretical relationship helpful in understanding the dynamic nature of stream erosion and valley development. A graded stream (according to the definition of J. H. Mackin, 1948) is one that has delicately adjusted its gradient so that over a long period of time (long enough to even out times of low flow and flooding), the velocity is just sufficient to move the load being delivered to the stream. A stream at grade is regarded as being self-regulating in the sense that its channel remains essentially stable or tends to regain the same form following any change in environmental conditions. The long profile of a graded stream channel generally shows a smoothly concave slope, steeper in the upstream reaches and flatter downstream. Irregularities in the long profile indicate that a stream is ungraded, with differential resistance of underlying rock units being the most likely cause. Wherever a sudden dip occurs in the long profile of a stream, the presence of a waterfall or rapids can be suspected. Although faulting, cliff retreat (as along a coast), or other circumstances can cause waterfalls, the most usual type is found where a stream crosses a contact between rocks of differing resistance and the weaker rock is being worn away more rapidly. The falls of the Potomac River at Washington, D. C., for example, are located at "the fall line" where the river crosses from crystalline rocks onto softer sedimentary rocks. As erosion continues, waterfalls are cut back headward (as at Niagara Falls) or downward, resulting in the formation of rapids. As falls are worn away, the stream approaches a graded condition. Eventually, as equilibrium is attained and the channel is stabilized, erosion and deposition become of less significance in the regimen of the stream.

SHAPE OF STREAM VALLEYS AND RETREAT OF SLOPES

As seen in cross-section, the most typical valley shape is that of a "V" although in many cases the point of the V tends to be more flaring than sharp. Where a stream has built a floodplain, the bottom of the valley is broad and flat. The most casual observation of hillsides suggests that slope angles and slope curvatures vary from very steep to very gentle and from convex through to concave. How can this infinite variety of valley shapes be explained, and how have these slopes come into being? It is obvious that as the channel of a stream is cut deeper, the valley increases in size, becoming wider as the slopes "retreat" backward into the bedrock or regolith (fig. 3.1). Weathering, mass wasting, and the processes of runoff are at work wearing the slopes away (Schumm and Mosley, 1973).

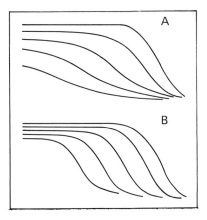

Figure 3.1. How valley side slopes retreat. **A:** Slope angles of the valley side become flatter as the slope retreats with the passage of time. **B:** As the slope retreats from the stream, with the passage of time, the slope angle of the valley side remains the same (parallel retreat of slopes).

It would seem logical to suppose that as a stream becomes graded and channel cutting slows down or ceases, the processes of valley widening continue to operate under the influence of gravity. Some theories have suggested that with passage of time, slope angles become flatter and the curvatures tend to change from convex toward concave. Field evidence to support this position is difficult to obtain, especially in areas where glaciation has affected the landscape. Some generalizations about slope retreat, developed from study of slopes in nonglaciated and arid or semiarid regions of the world, are summarized here and in the next chapter.

Thirty to 35 degrees is considered (with few exceptions) the maximum steepness of valley side slopes. In general, slopes are steeper on more resistant rocks. Where downcutting of a channel is very rapid, slopes are steeper than in situations where downcutting is minimal. Denuded slopes where talus material is being removed rapidly are steeper than slopes where debris accumulates and gradually buries much of the erosional surface. Slopes being cut in fine soil or silt tend to be steeper than those cut in loose sand or gravel. The lush vegetation of wet, tropical regions tends to inhibit slope erosion of valley walls causing them to be steeper, in general, than the slopes of valleys in humid temperate areas.

Slope angles are the result of interaction between local climate (including vegetation) and bedrock or regolith. The angle of a slope is also affected by the size of bedrock blocks that weather out, a factor related to jointing, character of bedding, lithology, etc. Slope angles are generally similar throughout local areas underlain by homogeneous rock.

Sometimes valley side slopes exhibit a series of segments sloping downward at different angles. Such slope segments may be the result of differential resistance of rock units, structural relationships, or rejuvenation of a stream's downcutting activities. Where long, gentle slope segments are found near the tops of the valley walls with steeper segments below, the association is referred to as "valley in valley." Probably some factor affecting the stream regimen resulted in a resumption of downcutting after the stream had developed a fairly wide valley with moderately gentle side slopes (Carson and Kirkby, 1972).

A conclusion that can be drawn from these observations is that valley shape and slope angles are the result of complex interactions involving type of rock, geologic structure, the elevation above base level or sea level, climate (including vegetation), and time. Whether slopes once formed persist in their inclination as they retreat, or whether slope angles decline as slopes retreat is an unsettled question (fig. 3.1). Do slopes eventually become graded in the sense that streams are graded? The philosophical aspects of these questions are discussed in Chapter 5, which deals with geomorphic systems.

ALLUVIATING STREAMS

Alluviation, or deposition by streams, occurs under conditions that are the converse of those producing erosion. Thus any circumstance causing the velocity of the stream to decrease brings about deposition (assuming that the stream is carrying a sediment load) (figs. 3.2, 3.5, 3.6). *Alluvium* is the general term for stream-deposited material. In many stream valleys, short-term conditions may be those favoring alluviation; but over the long run (geologic time), all alluvial material might be said to be in transit and at rest only temporarily. Normally a great deal of the mechanically transported load of a stream lies at rest for periods of time between episodes of movement. The well sorted sand, mud, and pebbles covering the floor of a channel make up the bed load of a stream.

Typical Alluvial Features

Bars are piles of alluvium built up near the middle of a stream bed. *Point bars* are deposits left on the inside of curves (fig. 3.3). Both types tend to be unstable. During times of flood, a stream overflows its banks and drops quantities of sand and silt on *natural levees* that are built up along the edges of the channel. These levees are seldom more than five

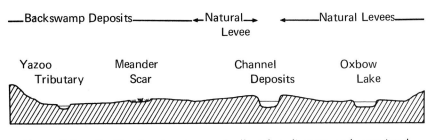

Figure 3.2. Profile showing types of alluvial sediments and associated topography on a floodplain. The coarsest deposits are in or close to the channel. The backswamp deposits contain the finest sediments.

feet high, are usually hundreds of feet wide, sloping away from the river. Beyond the levees, usually quite some distance from the main channel, lie the *backswamp deposits*, made up of silt and fine sand. Sometimes small streams flow down the natural levees away from the main river into the backswamps and low parts of the floodplain (fig. 3.4).

Floodplains

That part of the valley floor covered by water during floods is called the *floodplain* (fig. 3.2). As a river approaches a graded condition and the rate of downcutting decreases, the energy of the stream tends to shift to widening its channel rather than deepening it. The lateral migrations of the stream cut against the bases of the valley side slopes, increasing the rate of mass wasting and slope retreat and making the valley bottom ever broader, until the floor of the valley becomes many times wider than the stream channel. As the river migrates back and forth across the valley floor, its flow gradually planes off a flat erosional surface, called the *strath*. On top of this, during the shifting back and forth, the stream deposits irregular layers of alluvium (sand, silt, mud), whose deposition is related to the short-range and long-range fluctuations in the volume of the stream. During times of high water, additional deposition may take place as the river rises and spills out over the valley floor. In this manner floodplain surfaces are built as complex landforms resulting from the deposition, erosion, and reworking of alluvium by the river.

Floodplains are usually flat—seldom with a relief of over ten feet. Agriculture developed early on the fertile, easily arable, well watered bottom lands of the floodplains; and on them man's first experiments in urban living were made. Cities could be easily built on the flat surfaces, and the river was there for convenient transportation, communication, and the later development of trading and manufacturing. Throughout history, however, all these activities have been susceptible to interference from floods.

The Meandering Habit of Rivers on Floodplains

Many rivers, especially those flowing in wide, flat-floored valleys, have irregular channels that tend to develop into endless successions of wide loops or bends called *meanders*. A connected sequence of meanders forms a *meander belt*, which itself may wander in larger swings from one side of its valley floor to the other. Meanders develop or grow when streams flow at low gradients in soft material. Any irregularity in the channel deflects the main flow against first one bank and then the other, causing the meanders to shift. On the outside curves, the water flows faster, eroding the outside bank by undercutting and bank collapse. On the inside of the curves, the water slows down and deposition takes place. The combination of erosion on the outside and deposition on the inside of curves causes the whole meander to "migrate," or move laterally back and forth across the valley and in a downstream direction at the same time. The sweeping movement of the stream is often rapid. Bank erosion of many feet can occur in a few hours during times of high water. So extensive is the movement of the meanders that they intersect other parts of the channel and the meander cuts itself off. The main part of the stream rushes through the cutoff, while the former meander loop is left abandoned (fig. 3.3). Soon the old channel fills in at both ends of the loop and a long crescent lake is formed, called an *oxbow* (Scott, Arkansas, fig. 3.4). In shape, oxbow lakes tend to resemble the U-shaped frame or bow embracing an ox's neck.

Migrating rivers and meanders being cut off often create troublesome situations for floodplain inhabitants. When, through lack of foresight, or more likely an insufficient knowledge of stream dynamics, a political boundary is placed along a meandering river, that boundary is almost certain to shift—either gradually, with the passage of time, or suddenly, sometimes many miles in a few hours. Towns built on the banks of meandering rivers have been left miles away from river traffic by meander cutoff and have become "ghost towns" as a result.

Extensive engineering works have been carried on in an effort to prevent bank-cutting and channel-shifting, but usually this has been a losing battle. When one part of a river system is changed, a counter change develops nearby. Constant dredging, riprapping of banks, the placing of buoys and lights, and the use of other means of channel maintenance are necessary in order to permit year-round navigation.

The device of artificially straightening river meanders by digging a shorter, more direct channel down the middle of the valley has not been outstandingly successful where it has been tried. With the length of the channel reduced, the gradient becomes steeper and the velocity greater. The newly straightened river has a tendency to incise its bed deeply. When the tributaries follow suit, the whole valley floor becomes dissected and eroded.

The valley of the lower Mississippi River from Cairo, Illinois, southward is a classic example of a shifting river channel on a vast scale. With

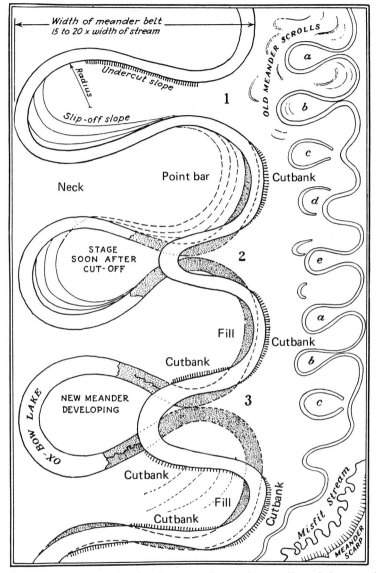

Figure 3.3. Stages in the development of meander cutoffs and the growth of new meanders. The main thrust of the flowing water erodes the outside of the curves, producing cutbanks. **1.** On the inside of the curves, where velocity is less, deposition of sediments builds point bars. **2.** The narrow neck between the cutbanks has been severed, and extensive deposition is going on in the slack water areas of the old meander channel. **3.** Further channel shifting has occurred, leaving the old channel now partly filled in as an oxbow lake. (After A. K. Lobeck, 1939, **Geomorphology,** McGraw-Hill; used with permission.)

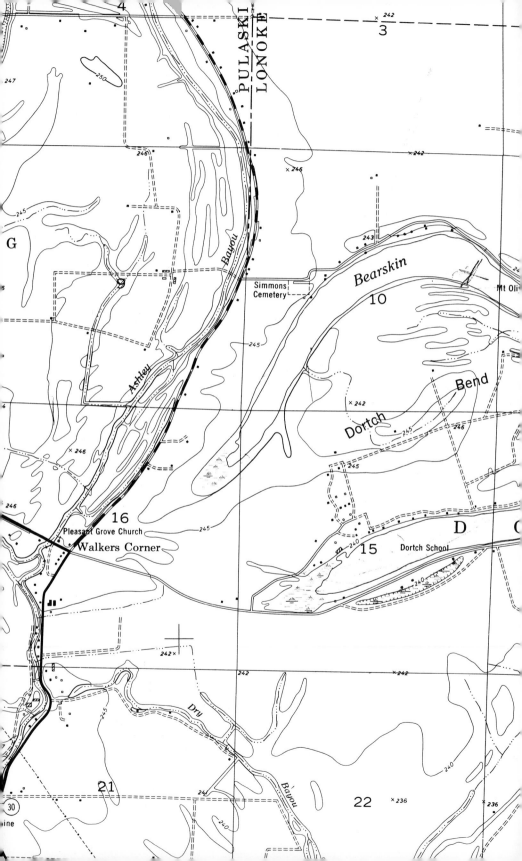

its great volume and large sediment load, the Mississippi has tremendous energy; and the magnitude of its changes is impressive. Man's efforts to maintain some semblance of control over the river are about a standoff. Sometimes the engineers are able to get the best of the river, but at other times the river overcomes them and changes its course or "goes on a rampage." With the passage of time, a great deal of trial and error, an accumulation of research data, and the expenditure of enormous sums of money, we are probably gaining a little in our attempt to control "Ol' Man River."

The geologic history of the lower Mississippi River helps to explain, to some degree, the complexity of the problem. The valley south of Cairo is not a true floodplain in the geomorphic sense, nor did it develop by lateral erosion in the manner described above. Beginning about 60 million years ago, marine sedimentation filled up the "Mississippi embayment"; during Pleistocene and Recent time, complex delta-building and floodplain accretion formed the present lower valley. Thus the thick wedge of marine sediments and the overall form of this part of the river valley are the result of geologic agents other than the present Mississippi River, which has, nevertheless, extensivly reworked the surface of the lower valley, producing features typical of an alluviating river.

Deltas and Fans

Deltas and fans are built up by streams when their carrying power is severely cut back, as when a stream changes gradient rapidly on coming out of a mountainous area onto a plain, or where a stream empties into a lake or the ocean. Where built on dry land, the triangular-shaped accumulations of deposits are called *alluvial fans* or *alluvial cones* (Valyermo, California, fig. 3.5).

Deltas are built out into the quieter water of a lake or bay by debris brought down from the areas of heavy erosion. As the main channel at the outlet becomes filled up, the flowing water is diverted, creating new distributaries, often in the pattern of a braided channel. Oscillation of flow among the distributaries causes deposition over a wide arc, which is how the characteristic fan or cone shape of these features is formed.

Figure 3.4. Scott, Arkansas (1:24,000, C.I. 5 ft., 1954). The Arkansas River is just to the west of the map area, which is a few miles south of Little Rock. Although the region is part of the lower Mississippi River Valley, the stream features shown were formed by the shifting of the Arkansas River rather than the Mississippi. Bearskin Lake is an oxbow. Ashley Bayou and Dry Bayou are meander scars. The sequence of events illustrated here has gone from cutoff channel to oxbow lake to oxbow-shaped swamps to depressions filled with sediment and organic debris and no longer holding water. These remain as meander scars on the valley floor. The sequence from cutoff to meander scar may take only a few hundred years.

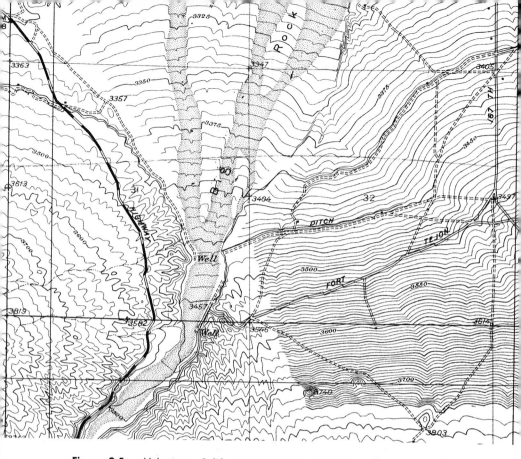

Figure 3.5. Valyermo, California (1:24,000, C.I. 5 ft. and 25 ft., 1940). Notice that this map has two contour intervals—a 25-foot one for the rugged mountains and foothills, and a 5-foot interval for the flatter topography of Antelope Valley. Big Rock Wash, an intermittent stream, flows down from the mountains and has spread an alluvial fan from the valley mouth, where the gradient drops sharply, out over the floor of Antelope Valley. The stream has shifted back and forth and has occupied several different channels (distributaries) on the fan.

As the outside edge or toe is built outward, more deposition takes place back at the apex in order to maintain flow. During high water, streams building deltas overtop their banks and fill in low places on the delta tops with alluvium. When flood waters recede, the main distributaries may be flowing in new, shorter channels to the sea. The delta at the mouth of the Mississippi River is extending itself by a complex pattern of growth of the natural levees along its distributaries. In order to keep the main channels open for shipping, engineers try to prevent overtopping and the development of new channels. The shape of the Mississippi delta has grown to resemble a bird's foot, with each distributary—or "pass," as they are known locally—rapidly growing longer. The appearance on a map of this type of river mouth is responsible for its picturesque name, *bird's-foot delta.*

Alluvial Terraces

A terrace is a landform that resembles a stairstep with a flat tread (the top) and a steep riser (the face). Terraces are of many types and are formed in a variety of ways, the most common being those found along the sides of stream valleys (fig. 3.6). Terraces represent a change in the regimen of the stream and are the result of its down-cutting into

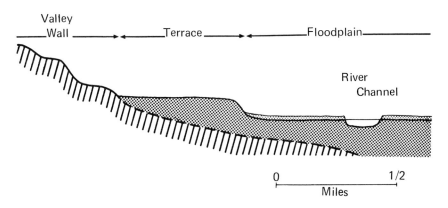

Figure 3.6. Profile of an alluvial terrace. When the valley was filled with alluvium, the floodplain surface was at the same elevation as the top of the present terrace. Due to changes in the stream regimen, the river stopped depositing and began to erode a new valley in the alluvium. The stream has spread a thin layer of alluvium on its new floodplain.

the thick blanket of alluvium spread over the valley floor. For alluvium to blanket a valley in a layer from a few tens of feet to a few hundreds of feet in thickness requires long continuing conditions favorable to alluviation. Eventually, because of factors such as a change in discharge, lowering of base level, etc., the river ceases to alluviate and becomes principally involved in eroding its bed by cutting a trench in the alluvium. If the river shifts laterally during the period of incision, the valley within the valley becomes wider than the channel and a small floodplain is formed. The tops of the alluvial terraces, built up by deposition during conditions of overbank flow, may have meander scars and other features typical of floodplains. Terrace faces may show a scrolled pattern where meanders have migrated up against the slope.

Not all terraces, it should be noted, are produced by the erosion of alluvium. Some are made of bedrock, and others are formed by faulting. In rare cases, alluvial terraces may develop from the erosional effect of a swinging meander on an alluvial fan.

SUMMARY

The hydrologic cycle, powered by the sun's energy, supplies water for the processes of erosion. Stream erosion (the picking up and transporting of material) is accomplished by hydraulic action, abrasion, and solution. Streams transport material by traction, saltation, suspension, and solution. Factors controlling stream erosion are velocity, volume, gradient, cross-sectional area, and load, with velocity being the most important. Downward erosion by streams is limited by base level. Streams work toward a condition of equilibrium between gradient and load (graded stream concept). Slope angles and valley shapes are the result of interactions involving rock type, geologic structure, relief, climate, and time.

Examples of Landforms Produced by Stream Action in Valleys and Channels

Erosional: valleys, divides, ridges (interfluves), channels, meanders, oxbow lakes, cutbanks, meander scars, meander scrolls, waterfalls, straths.

Depositional: channel bars, point bars, floodplains, natural levees, backswamp deposits, deltas, alluvial fans and cones, alluvial terraces (also partly erosional).

Chapter **4**

The Development of Stream-Eroded Landscapes

TOPICS

The development of drainage systems
Relief and drainage texture
Morphometric terrain analysis
Drainage density and basin characteristics
Classification of streams and valleys
Drainage patterns
Integration and adjustment of drainage
Misfit streams, incised meanders
The influence of geologic structure on landscapes
Genetic types of mountains
Stream erosion and deposition in desert regions
Development of pediments
Regional drainage in arid regions
Continental sedimentary deposits

Landscapes sculptured by streams are most typically part of large, integrated, and coordinated systems. The Mississippi-Missouri-Ohio drainage system and the Amazon basin, for example, each cover half a continent. How can one explain the development of such systems? To say that "from little trickles mighty rivers grow" is to give a superficial answer. The evolution of a stream system is an integral part of the descriptive geologic history of an area and involves climatic factors, the ruling influence of structure, and modifying aspects of tectonic uplift or downwarp. Beyond the satisfactions of seeking out these scientific answers, is a growing need for greater understanding of floods, pollution, water supply, conservation, and the like, as these relate to river systems. If environmental quality is to be improved or maintained, dynamic and pragmatic approaches need to be worked out that are in harmony with natural forces.

THE DEVELOPMENT OF DRAINAGE SYSTEMS

During and after precipitation, water that does not evaporate or infiltrate becomes runoff, draining the land surface as *overland flow* and as *stream flow*. In always seeking to reach lower levels as efficiently as possible, runoff organizes itself into drainage systems that tend to have similar patterns and hydraulic relationships. Overland flow may take the form of *sheet flow* that drains as a thin film over smooth rock or as a scarcely visible network through grass, leaves, etc. Where small bumps and hollows occur, sheet flow is diverted and becomes concentrated in tiny rills that erode parallel runnels or shoestring gullies, especially in soil unprotected by a cover of vegetation. In times of high runoff, small gullies enlarge as streamlets overtop their sides and join flow with larger, adjacent gullies. Shallow rills, diverted to deeper rills, flow at an angle to, rather than parallel to, the original slope. This process is called *cross-grading*.

Where overland flow converges on a slope to form a streamlet, the increase in volume results in increased erosion at that point, forming a short, steep pitch that is eroded headward. During heavy runoff, this process of *headward erosion* enables a small stream to cut back and lengthen its gully rapidly. The gully cannot usually be eroded all the way to the top of the slope, however, because the head of the stream reaches a point where the collecting area is insufficient for overland flow to produce enough water for further back-cutting. In the development of a drainage network, the units of length of the overland flow are quantitatively related to other parameters of the system. On any given land surface, the minimum or critical length of overland flow required to produce runoff sufficient to cause erosion depends upon the slope, the intensity of runoff, the degree of infiltration possible, and the resistance of the soil or surface to erosion.

The area that each stream, large or small, is able to drain is called its *drainage basin*. Within a basin, headward erosion is limited by a *drainage divide* or crest, from which runoff goes down on either side. Similarly, a stream valley cannot widen itself indefinitely because a basin divide (interfluve) exists between each valley side slope and the nearer slope of the adjacent stream. The channel bed itself can be eroded downward only to base level. Thus controls limit each type of growth. A drainage system can be visualized as a funneling mechanism that efficiently collects the various types of runoff in the higher levels and draws them out through progressively lower levels toward a somewhat constricted exit, which is the channel of the main stream. The outline of a drainage system is often pictured as pear-shaped, with the pear stem representing the major stream leaving its basin.

Relief

Vertical dimensions are useful in describing and comparing drainage systems. Maximum relief (the vertical distance between highest and lowest points) can be calculated easily for a quadrangle, a state, a continent, etc. *Drainage relief,* a more meaningful parameter, is a determination of the vertical distance between a stream and an adjacent interfluve. From numerous values of drainage relief, average relief for an area can be calculated. In ordinary usage, relief of less than 250 feet is considered low; from 250 to 500 feet, medium; and greater than 500 feet, high.

In order to describe a drainage system in more detail, combinations of measurements can be made, such as stream gradients and valley side slope angles with cross-sectional shape descriptions and ruggedness numbers. (One example of a ruggedness number is derived by dividing the vertical distance or relief by the horizontal distance between the two points used in calculating the relief). Another useful approach is to construct a curve to show the percentage of an area that lies above and/or below a certain elevation. This kind of area-altitude diagram is called a *hypsometric curve* (fig. 4.1).

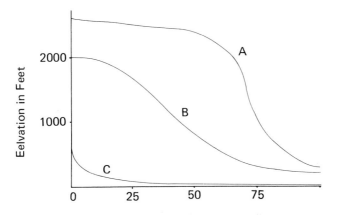

Figure 4.1. Hypsometric curves for three different landscapes. **A.** A region with most elevations above 2,000 feet. **B.** A region having a small area almost 2,000 feet high with the rest of the terrain largely in slope. **C.** A region with a small area nearly 500 feet high, with the rest being low flatlands. In terms of the Davisian cycle, **A** can be described as being in a stage of youth, **B** as maturity, and **C** as old age. (See Chapter 5.)

Drainage Texture

Landscapes sculptured by streams can be described as having a close (dense) texture or a wide (coarse) texture. This is a means of pointing out the degree of proximity of individual streams and their drainage basins. (Compare Waipio, Hawaii, fig. 1.3, p. 7, with Soda Canyon, Colorado, fig. 4.9, p. 58). Initial spacing of streams may determine the maximum relief that can develop in a region. If streams are closely spaced while downcutting proceeds, the valley side slopes intersect in the interfluves. But as the valleys are deepened farther, the relief stays about the same because the divide is lowered, also. In general, relief is higher and drainage coarser in regions of massive crystalline rocks than in areas of mixed rock types or in regions underlain by sedimentary rocks. The reason for this is probably that massive crystalline rocks tend to occur in high mountain masses. Moreover, erosional slope angles are somewhat steeper when cut in massive rocks.

MORPHOMETRIC OR QUANTITATIVE TERRAIN ANALYSIS

Sciences such as climatology and hydrology have, for many years, dealt with the collection and collation of vast amounts of data related to the atmosphere, rainfall, and stream flow. Statistical methods were developed for the utilization of these data, especially in the areas of predicting runoff, infiltration of ground water, and the fluctuations and sizes of stream flow. Since the shape and size of landforms are important in such studies and predictions, it became necessary to develop terrain descriptions utilizing appropriate quantitative parameters.

Robert E. Horton (1945) ,a hydraulic engineer, worked out a number of basic principles for the dimensional analysis of streams and drainage basins. The various types of measurements may be divided into three groups as follows (modified from a summary by Strahler, 1964): (1) linear properties that are measured or counted solely from channel networks and basin outlines; (2) properties based on areal measurements; (3) properties based on vertical differences (elevation and relief).

Stream ordering is a method for indicating the complexity of a fluvial system by assigning each stream segment a number appropriate to its rank and location. First-order stream are single and unbranched. When two first-order streams join, they form a stream of the second order. The joining of second-order streams creates a stream of the third order, and so on. If, however, a first-order stream enters a stream of the second order (or any higher order), no new order is formed. Small rills and streamlets (that may or may not be intermittent) are first-order streams; most small creeks and brooks are third- or fourth-order streams; the trunk stream of a watershed always has the highest number in a fluvial system. The lower Mississippi River, for example, has been calculated to be about a tenth-order stream.

After the ordering of a drainage network has been completed, counting the number of streams of any one order and comparing the total with the total number of tributaries of the next higher order brings out a mathematical relationship called the *bifurcation ratio*. This is a proportion that tends to remain constant within a system between one order and the next. Usually about three to five times as many streams are found in any one order as can be counted in the next higher order. Horton's (1945) *law of stream numbers* expresses this principle: The numbers of stream segments of successively lower orders in any drainage basin tend to approximate a direct geometric series, beginning with a single stream segment of the highest order and increasing according to a constant bifurcation ratio.

Order	Number of Segments	Bifurcation Ratio	Average Length of Segments (miles)	Length Ratio	Average Watershed Area (sq. miles)	Average Channel Slope (percent)
1	5,966	........3.9	0.09	........3.3	0.05	0.185
2	1,529	........4.0	0.3	........2.7	0.15	0.091
3	378	........5.7	0.8	........3.1	0.86	0.054
4	68	........5.3	2.5	........2.8	6.1	0.022
5	13	........4.3	7	........2.9	34	0.088
6	3	........3.0	20		242	0.0013
7	1		8+ (not complete)		550 (not complete)	0.0002

Figure 4.2. Allegheny River drainage basin characteristics (after Morisawa, 1959).

Horton also formulated a *law of stream lengths* based on a value called the *length ratio* that tends to be constant within a given fluvial system. First-order stream segments in a system tend to have the shortest lengths, while mean lengths of streams in the higher orders are found to increase by a ratio of about three with each order, going from lowest to highest. The law of stream lengths (as modified by Strahler, 1969) is stated as follows: "The cumulative mean lengths of stream segments of successive orders tend to form a geometric series beginning with the mean length of the first-order segments and increasing according to a constant length ratio."

The mathematical relationships that have been demonstrated by the law of stream numbers and the law of stream lengths appear to exist also in many watersheds between order of streams and mean basin areas, order and gradients, and order and relief of basins. Formulations of these relationships come under the general heading of "laws of drainage composition." In spite of inconsistencies that exist, quantitative investigations of basins and stream networks continue to suggest an orderly development. Figure 4.2 presents a compilation of data taken on the Allegheny River system, which drains portions of the Allegheny plateau and the Appalachian Mountains (Morisawa, 1959). Even without being plotted on a graph, the geometric relationships are apparent.

In another type of compilation, Leopold, Wolman, and Miller (1964), using ratios worked out on small drainage basins, projected the resulting totals to cover the entire United States (fig. 4.3). These predictions are based on the degree of accuracy that can be obtained from using 1:62,500 scale maps. For each increasing stream order in this tabulation, the number of streams decreases by approximately four-fifths, the average length doubles, the total length decreases by one-half, and the average drainage area increases fivefold.

Order	Number	Average Length (miles)	Total Length (miles)	Mean Drainage Area, Including Tributaries	River Representative of Each Size
1	1,570,000	1	1,570,000	1	
2	350,000	2.3	810,000	4.7	
3	80,000	5.3	420,000	23	
4	18,000	12	220,000	109	
5	4,200	28	116,000	518	
6	950	64	61,000	2,460	
7	200	147	30,000	11,700	Allegheny
8	41	338	14,000	55,600	Gila
9	8	777	6,200	264,000	Columbia
10	1	1,800	1,800	1,250,000	Mississippi
Total			3,250,000 (approx.)		

Figure 4.3. Number and length of river channels of various sizes in the United States (excluding tributaries of smaller order) (after Leopold, Wolman, and Miller, 1964).

Drainage Density and Stream Frequency

In order to express quantitatively the degree to which a watershed is well drained or dissected, the *drainage density* and *stream frequency* may be measured. To obtain a value for drainage density, one divides the total length of all streams in a drainage basin by the area of the basin, being sure to use similar units of measurement; i.e., total linear miles by total square miles. Stream frequency is the number of channels per unit area; that is, the number of streams in a basin divided by the basin area. Drainage density and stream frequency are quantitative expressions of drainage texture. Study of many watersheds has shown that drainage densities of 3.0 to 4.0 miles of stream channel per square mile of area are low. A region of low drainage density has a coarse texture with gross or massive topographic features. A region of high drainage density—in the range of densities of 30 to 40 miles of stream channel per square mile—can be described as having fine texture. Climatic factors and the physical characteristics of a region influence drainage density and stream frequency to a considerable degree.

Basin Characteristics

Quantitative relationships derived from data on basin shape and slope characteristics are less clear cut than those worked out from linear measurements but do suggest that some ratios may have practical implications. Large fluvial systems tend to have longer rivers with gentler gradients and fewer streams of the first order. Basins and divides with relatively steep valley side slopes tend to have fine texture, high relief, and tributaries with steep gradients. Valley side slopes are directly related to channel slopes; i.e., high capacity streams are necessary to carry away large volumes of coarse debris coming down steep valley walls. More gentle, smoother valley sides tend to contribute only fine debris and the stream channel below develops a flatter gradient. Inasmuch as variables such as valley width, valley depth, valley cross-sectional area, etc., vary proportionally (as power functions) with stream discharge, data derived from these parameters can be used to work out prediction equations. The observation has been made that individual parts of drainage systems are related proportionally, and that stream networks may have evolved according to the *law of allometric growth,* which was developed from the study of biologic organisms. This law states that the relative rate of growth of an organ is a constant fraction of the relative rate of growth of the whole plant or animal (Strahler, 1969).

Although the fundamental causes of the quantitative relationships discussed in this section are not completely understood and much more study of drainage basins needs to be done, the principles and theories so far derived from collected data do suggest the presence of similar controlling physical factors in the development of nearly all drainage systems. When a parameter or ratio for a particular stream system is not in agreement with the expected value, almost invariably a geologic reason can be found for the anomaly.

If an investigator is concerned about a particular aspect of a landscape or watershed, he can devise a special "measuring parameter." The number of measurable or especially derived parameters that might be developed is virtually limitless. In all cases, however, "standard" values must be collected in common units and in a prescribed manner so that uniformity is approximated. Caution needs to be used, also, whenever parameters are employed for descriptive and comparative purposes to insure that the procedures do not contradict actual field evidence. Many impossible situations have been "proved" by means of statistical data fed into electronic computers. A favorite illustration among geomorphologists (long antedating the age of computers) is Mark Twain's attempt at "quantitative analysis" of the lower Mississippi River. Twain was fascinated by the phenomena of meander cutoffs and "river shortening." In *Life on the Mississippi,* he tells about cutoff after cutoff that had occurred along the river and how much shortening of the river had resulted each time. Then he concluded:

> . . . In the space of 176 years the lower Mississippi has shortened itself 242 miles. That is an average of a trifle over 1-1/3 miles per year. Therefore, any calm person who is not blind or idiotic can see that . . . just a million years ago . . . the lower Mississippi River was upwards of 1,300,000 miles long and stuck out over the Gulf of Mexico like a fishing

rod. And by the same token any person can see that 742 years from now the lower Mississippi will be only 1¾ miles long and Cairo [Illinois] and New Orleans will have joined their streets together. . . . There is something fascinating about science. One gets such a wholesale return of conjecture out of such a trifling investment of fact!

CLASSIFICATION OF STREAMS AND VALLEYS

Both genetic and descriptive terms are used to explain reasons for a stream's location, its direction of flow, its pattern, and the shape of its valley. Appropriate use of such classifers helps to identify topographic features quickly and clearly. The key relationship looked at in the classification of streams is the causal factor controlling the direction of flow. When the flow is in accord with the regional slope, a stream is described as *consequent;* i.e., the stream flow is "in consequence of the slope." Where geologic structure is the controlling factor, the streams are classified as *subsequent* because the flow is subsequent to the underlying structure. Types of subsequent streams and valleys are further classified by the following terms:

Strike, homoclinal, or monoclinal—parallel to the strike of dipping beds.
Anticlinal—parallel to the axis of an anticlinal fold.
Synclinal—parallel to the axis of a synclinal fold.
Fault—parallel to the outcrop of a fault.
Joint—parallel to bedrock joints. (Streams may follow a rectangular pattern under the control of more than one set of joints).

Where several tributary streams are controlled by structure, a special nomenclature is used (fig. 4.4). These tributaries would, of course, all be called subsequent and would normally flow into a major stream.

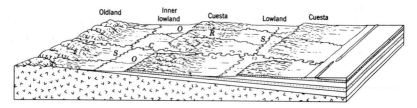

Figure 4.4. Types of streams and their relationship to structure. On this coastal plain, where sedimentary rocks dip gently seaward, stream types are indicated as follows: **S** = subsequent; **C** = consequent; **O** = obsequent; **R** = resequent. Cuestas have developed on the more resistant beds. Drainage follows a trellis (rectangular) pattern. (From A. K. Lobeck in A. N. Strahler, 1973, **Introduction to Physical Geography,** 3rd ed., John Wiley and Sons. Copyright ⓒ 1973.)

Two types of streams may flow into the subsequent streams; namely, the right and left bank tributaries of these subsequent strike streams. Those flowing in the same direction as the principal stream are called *resequents* (recent consequents), and those flowing in the opposite direction from the major stream are *obsequents* (opposite consequents). To fully round out the classification of "-quent" streams, the term *insequent* designates streams whose direction of flow is not controlled by any recognizable factor! (Streams that might be thought of as insequent are the new streams developing on the uneven surfaces of glacial drift plains).

Genetic terms are applied to stream valleys in situations where up or down movement of the earth's crust has influenced valley shape. *Drowned valleys* have been submerged and have become occupied by the sea. A fiord, for example, is a glacial valley that has been drowned. *Rejuvenated valleys* are those that have been uplifted, with the result that their floors are being cut deeply by their streams.

Rivers flowing across the trend of mountain masses (that one might suppose would be a barrier to stream development) are called *transverse streams*. Well known examples are the Delaware Water Gap between Pennsylvania and New Jersey, the Potomac River at Harpers Ferry in Virginia, the Arkansas River at the Royal Gorge in Colorado, and the Columbia River in the Columbia Gorge where it crosses the Cascade Range between Washington and Oregon (fig. 4.5).

Two explanations, in terms of their genesis, have been proposed for transverse streams. An *antecedent stream* is one whose direction of flow within a valley was established before the mountainous structure was uplifted (fig. 4.6). Apparently antecedent streams were able to erode rapidly enough to maintain their course across the structure as it rose. A *superimposed stream* is one whose valley and direction of flow were developed much later than the structure (fig. 4.7). In the process of deepening its valley, the stream's course has cut across the buried structure, upon which it has become superimposed. In principle these two concepts are simple enough; but in nature, considerable field data and a well developed investigation of the geologic history of a region are needed in order to determine whether a transverse stream is antecedent or superimposed. In most situations that have been carefully studied, the evidence indicates that the streams have been superimposed.

Some so-called valleys (similar in their negative topographic form to true valleys) have been formed by geologic activity other than stream erosion. Most of these features are the result of structural or tectonic movement. Folding produced the Great Valley of California. Faulting has resulted in rift valleys and grabens in various parts of the world; examples are the Jordan River valley and the African rift valleys. Collapsed valleys (uvalas) are characteristic of karst topography (Chapter 7, p. 133). Many valleys have a composite origin. A stream valley may have been widened into a glacial trough as, for example, the Hudson River valley. Structural features, such as Paradox Valley in Colorado, have

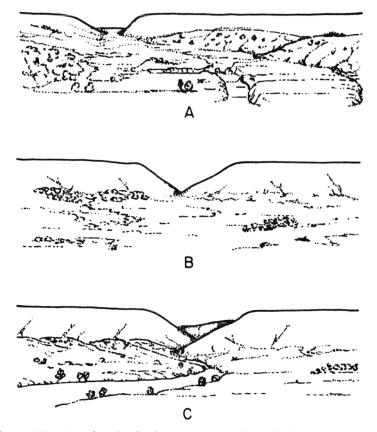

Figure 4.5. Accordant levels along mountain ridges (hogbacks of resistant rocks) and water gaps. **A.** Delaware Water Gap. **B.** Susquehanna River water gap near Harrisburg, Pennsylvania. **C.** Water gap near Cumberland, Maryland. (From **Physiography of the United States** by Charles B. Hunt. W. H. Freeman and Company. Copyright © 1967.)

been modified subsequently by streams. A terrestrial valley unaffected by stream erosion probably does not exist on this planet.

Drainage Patterns

Stream systems may be classified by their geographic or map patterns, as well as by their genetic development. Usually the pattern of a system involves a main stream and its tributaries within their drainage basins. *Drainage patterns* are viewed from the perspective of an imaginary observer high above the earth—a map view, in other words.

Basically patterns are of two types: dendritic and trellis, corresponding roughly to the genetic classifications of consequent and subsequent; that is, slope controlled or structurally controlled. A *dendritic drainage*

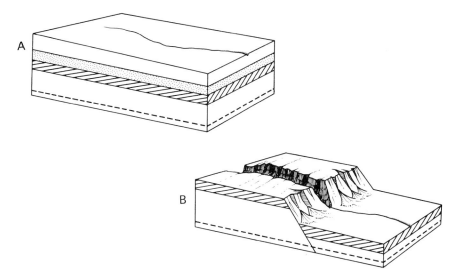

Figure 4.6. Evolution of an antecedent stream. In block **A**, a stream has established its course across the gently sloping surface. Block **B** shows that although faulting has occurred, the stream has been able to maintain its course across the upthrown block. Passage of a considerable length of time is assumed between the stages of block **A** and block **B**. Sea level is represented by the dashed line.

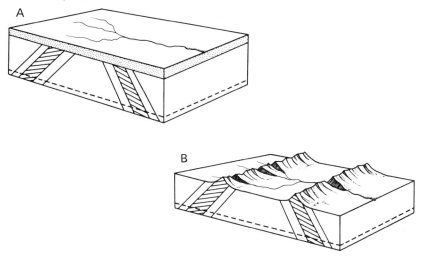

Figure 4.7. Evolution of a superimposed stream. In block **A**, a stream is cutting down on a horizontal stratum. In block **B**, long continued downcutting has exposed the buried structure. The stream now flows across the resistant, ridge-forming beds. The dashed line on each block represents sea level.

pattern, made up of many consequent streams, resembles the trunk and limbs of a great oak, with large branches forking into smaller ones that, in turn, successively divide into a multitude of twigs. Tributary streams usually join main streams at acute angles pointing downstream—this being an indication of slope control (Soda Canyon, Colorado, fig. 4.9, and Casey, Illinois, (fig. 5.3, p. 76). Significant types of dendritic drainage patterns include the following:

Radial—outward-flowing streams, as on a volcano (Mount Rainier National Park, (fig. 6.1, p. 95).

Centripetal—inward-flowing streams, such as those found in basins of interior drainage.

Anastomosing—a braided pattern of stream channels, joining, separating, and rejoining (Valyermo, California, fig. 3.5, p. 42).

Parallel—a large number of streams flowing down a steep slope without the development of many tributaries (Waipio, Hawaii, fig. 1.3, p. 7).

Trellis or *rectangular drainage patterns,* being made up of both consequent and subsequent streams (fig. 4.4) are characterized by parallel or rectangular systems of main and tributary streams that reflect both the slopes and the underlying structures. Trellis patterns tend to have paired tributaries that usually join the main stream at right (rather than acute) angles, indicating structural control. In rectangular patterns, the main stream and tributaries have straight reaches and right angle bends but the tributaries are not usually paired (fig. 4.8).

Indications of the nature of the underlying rocks and their structure can be derived from drainage patterns. Slope-controlled patterns suggest that the bedrock is not influencing the development of the drainage. We may conclude that the rock is either homogenous throughout or made up of flat-lying beds. Trellis and rectangular drainage patterns, being structurally controlled, delineate the size, shape, and location of the underlying structures (fig. 4.11).

Integration and Adjustment of Drainage

Fluvial systems, both in their components and in respect to adjacent systems, tend to establish and maintain equilibrium. Regional drainage is said to be *integrated* when the upper portions of several unconnected drainage systems have grown headward as far as they can, producing narrow divides, and when the collecting areas of the basins are unable to furnish the additional water that would be needed for further elongation and elaboration of the stream networks. Equilibrium of a watershed is dynamic in the sense that morphologic characteristics of the basins are constantly changing, but always in the direction of balancing gains and losses. Some tributaries may be lengthening, but others are being shortened. A drought or a drop in the water table results in the temporary or permanent loss of some tributaries. (Compare the degrees of

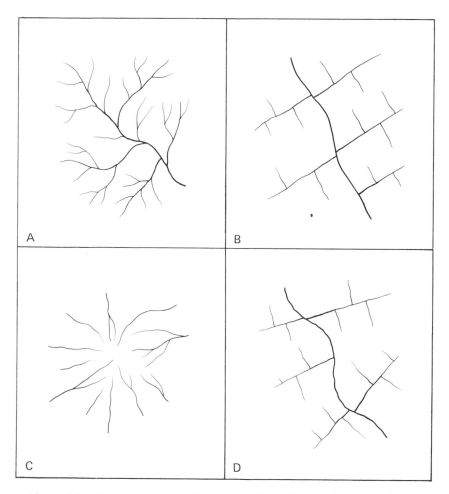

Figure 4.8. Drainage patterns: **A.** dendritic. **B.** trellis. **C.** radial. **D.** modified trellis.

integration shown in Boone, Iowa, fig. 6.5, p. 107; Casey, Illinois, fig. 5.3, p. 76; Soda Canyon, Colorado, fig. 4.9; and Draper, Virginia, fig. 5.4, p. 78).

In areas where drainage systems are developing, some streams or rills, because of larger volume, steeper gradients, or other favorable circumstances, are able to erode headward faster and thus divert or *capture* the flowing waters of nearby tributaries. Capture, or stream piracy, is particularly evident in regions where the direction of stream flow is largely controlled by weakness in the rocks or by other structural factors. The subsequent tributaries that drain over the less resistant rocks grow headward faster and capture other streams. Eventually, with domi-

nance of streams favored by structure, *adjustment of drainage* is established. A trellis or rectangular drainage pattern indicates that adjustment of drainage to structure has taken place.

Misfit Streams and Incised Meanders

Where map and field data indicate the existence of misfit streams or incised meanders in a drainage area, these can be used in interpretations of regional geologic history. *Misfit streams* suggest that climatic changes have occurred in the recent geologic past; *incised meanders* may have been caused by regional uplift.

In relation to their present valleys, some rivers appear to have been far larger in the past than they are at the present time. Such streams are called misfit, although a more accurate designation might be "underfit." However, factors other than volume, such as rock structure and resistance, may have played an important role in determining valley widths and meander patterns in those regions where misfit streams are found. G. H. Dury (1960) has suggested that most streams in the humid temperate zones are moderately misfit because runoff decreased in the recent past (postglacial time) as the large volume of glacial meltwater—which both eroded and deposited along the valleys—diminished in the warming climate (Palmyra, New York, fig. 6.7, p. 110).

The Minnesota River (which joins the Mississippi River at Minneapolis) is a misfit stream that once served as an outlet for glacial Lake Agassiz. About 6,000 to 10,000 years ago, Lake Agassiz was an ice-dammed body of water occuping the valley of the Red River (of the North) between Minnesota and North Dakota. Numerous streams that cross Illinois and Indiana (e.g., the Illinois and Wabash Rivers) drained various stages of the Great Lakes in glacial and postglacial times and now show evidence of being underfit.

Dry stream beds (arroyos) in arid and semiarid regions are not thought of as misfit because they are "in adjustment" from a long-range point of view to infrequent but sometimes high discharge. Nevertheless, it is possible that rainfall and runoff have changed more in some arid regions since the Pleistocene than in humid regions.

Incised meanders are found on streams with steep and irregular gradients; and yet the streams are flowing in meandering patterns more

Figure 4.9. Soda Canyon, Colorado (1:62,500, C.I. 50 ft., 1912). In this area, which has characteristics typical of Davisian youth (Chapter 5), a dendritic drainage pattern is being developed in a plateau region where a thick sequence of flat-lying rocks has been uplifted to a moderately high elevation. Interfluve areas fit the descriptive image of "broad and flat divides." The Mancos River has slowed its downcutting and is widening its valley. Streams are few in the area and some are intermittent. This suggests that the climate is semiarid.

typical of rivers that are approaching a graded condition (Oolitic, Indiana, fig. 7.6, p. 133). When seen from above, or as depicted on a map, incised meanders have the characteristic looping patterns usually seen on a well developed floodplain (fig. 4.10). Examination in the field, however, shows that the loops and bends are cut down (sometimes quite deeply) into bedrock, with steep valley walls rising on both sides of the stream. It has been suggested that incised meanders, especially where valley side slopes are relatively symmetrical, were formed by the uplifting of the land, causing the river to cut downward while maintaining its meandering pattern. Where the valley side slopes are not symmetrical but tend to be steeper on the outside and gentler on the inside of meanders, they may have been formed by the stream's "slipping off" while downcutting. Incised meanders of this type would not be an indicator of uplift. Although examples of both types exist in nature and have been studied, the differences between them are subtle. Interpretations suggesting uplift, therefore, must be made cautiously.

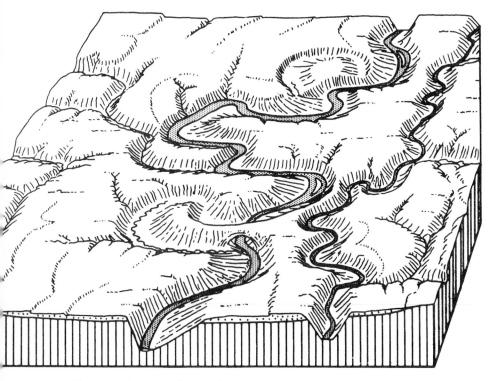

Figure 4.10. Incised meanders on two streams of different size. (After E. Raisz in A. N. Strahler, 1969, **Physical Geography**, 3rd ed., John Wiley and Sons. Copyright © 1969.)

THE INFLUENCE OF GEOLOGIC STRUCTURE ON LANDSCAPES

Global tectonics determine the presence and massive shape of mountain systems, but as erosional processes have worked along through millions of years, different stream-eroded landscapes have evolved. For this variety, the dissimilar structures and rock units of the mountain systems are largely responsible. The sizes and shapes of structural units define the major divides, delimit the overall drainage basins, control the direction of flow of principal rivers, and, in general, provide a framework within which erosional processes take place (fig. 4.11).

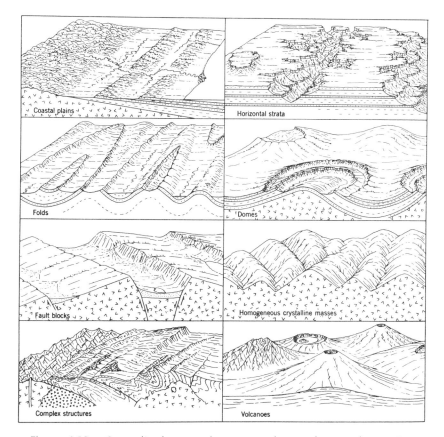

Figure 4.11. Generalized types of structures that produce, under erosion, different types of landscapes. The influence of structure is apparent in the topographic forms shown here. (From A. N. Strahler, 1973, **Introduction to Physical Geography,** 3rd ed., John Wiley and Sons. Copyright © 1973.)

Plains

Plains are areas of low relief and low elevation underlain by flat-lying rocks (Atlantic and Gulf coastal plain) or homogeneous rocks. Structure has little effect because drainage is slope controlled. Uplands are gradually transformed into a dendritic pattern of low, rounded ridges separating streams. Changes in rock types are reflected in the development of *cuestas*, which are gently sloping plains bounded on one edge by an escarpment. On the Gulf coastal plain of the United States in Texas and Alabama, cuestas face inward with gentle backslopes inclined downward toward the Gulf of Mexico (fig. 4.4).

Plateaus

Plateaus are areas of low relief at high elevations underlain by flat-lying rocks (the Colorado plateau) or homogeneous rocks. On dissected plateaus (which have high relief), the structure shows up in alternating slope angles on the sides of deep valleys and flat interfluves underlain by resistant beds. Drainage is slope controlled. Depending on the amount of rainfall, the landscape may be angular, as in semiarid regions, or more rounded or smooth, as in humid climates (fig. 4.9).

Fold Mountains and Dome Mountains

Some mountain landscapes are formed by the erosion of a complex series of folds (anticlines and synclines, usually doubly plunging). Drainage patterns are directly influenced by outcrop patterns and the location of dipping beds. The more resistant edges of formations (rock units) stick up as hogbacks that occasionally hook and loop across the landscape as erosion attacks the plunging noses of the folds (fig. 4.12). Subsequent streams follow outcrop patterns of the less resistant rocks; transverse streams cross hogback ridges through water gaps. Typical landform assemblages that have evolved by stream erosion on folded rocks are described as *valley and ridge topography*.

Geologic domes (fig. 1.1, p. 3) and basins develop circular patterns of streams and ridges rather than linear patterns, such as those found on fold belts. The topography of northeastern France, as well as the Weald district of southern England, developed on a large, shallow geologic basin. Several encircling outfacing cuestas have been formed there by stream erosion.

Fault Mountains

Faults are fractures in the earth's crust along which movement has occurred. Streams that erode headward or follow along the traces of faults are subsequent. Bright Angel Creek (a tributary of the Colorado River in the Grand Canyon) has eroded a straight valley along a vertical fault that extends many miles back into the Kaibab plateau, which forms

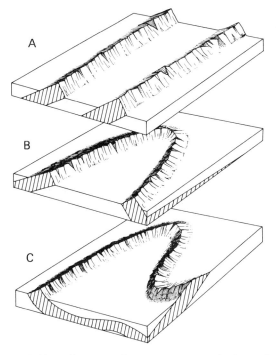

Figure 4.12. Hogbacks produced by differential erosion of dipping units of sedimentary rock. Block **A**: Resistant beds dip toward the left and strike from lower left to upper right. The gentler or dip slopes face away from the reader. To determine the direction of dip of hogback-forming strata, assume that they dip in the downhill direction of the more gently sloping side of the hogback. Block **B**: The erosion of a resistant anticlinal fold that is plunging toward the upper right has formed a hogback that loops. Block **C**: Zigzag hogbacks, on folded sedimentary beds tend to loop in broad curves across the axes of synclines and form sharp prows when crossing the axes of anticlines. (An axis is a theoretical line or "ridgepole" along the top of a folded bed. A fold is plunging when its axis is not horizontal). In this block both the syncline (on the left) and the adjoining anticline (on the right) are plunging toward the lower left.

the north rim of the Grand Canyon. Faults frequently occur in more or less parallel sets and the areas of rock in between these faults may move up or down, relative to adjacent blocks. Thus faulting may produce upfaulted mountain blocks between lower standing blocks. Usually the sides of such fault block mountains tend to be straight or linear. Faults may lie on both sides of such mountains or on one side only. Large mountain masses bounded by faults are the Teton range in Wyoming, the Wasatch range in Utah (east of Salt Lake City), and the Sierra Nevada of California and Nevada. Smaller block fault ranges, ten miles or so wide, separated by alluvium-filled basins, are found in Nevada and Utah. These structures, roughly rectangular in shape, are referred to as basin and range and the resulting landscapes are described as *basin and range topography*. The geologic structure and lithology of the constituent rocks are usually complex mixtures of folded and faulted sedimentary and volcanic rocks. Erosion reduces the upfaulted blocks and the sediments produced are deposited in adjacent basins. Where rocks of differing resistance to erosion occur adjacent to each other, distinct peak ridges and valleys can be seen. Drainage is radial off the mountains and, in arid and semiarid regions, centripetal into the basins (interior drainage) (fig. 4.8).

Volcanic Mountains

Most volcanic mountains are accumulations of pyroclastic (volcanic) rocks interbedded with cooled lavas, such as Mt. Rainier (fig. 6.1, p. 95) and Mt. Hood. Drainage develops radially outward from high areas and differing resistance of the various kinds of volcanic rock becomes evident as weathering and erosion work away. At Spanish Peaks, Colorado, for example, the filled and hardened throats and lava conduits of eroded volcanoes now stand as resistant remnants (see pp. 136-8).

Erosional Mountains

In some cases, such as the Catskill Mountains of New York and some highlands near the Grand Canyon, flat-lying (or homogeneous) rocks have been uplifted (forming plateaus) and then have become deeply eroded by streams into very diversely shaped mountains. In this topographic situation structure is not a controlling factor.

Complex Mountains

As seen in nature, most mountains cannot be classified by simple genetic type according to the outline above. Sometimes mountains have both folds and faults, with sedimentary rocks cut by masses of plutonic rocks or volcanic rocks and, in varying degrees, affected by metamorphism. Complex mountains may have been uplifted so that erosion has many thousands of feet of rocks to work over and remove. Within limited areas, topography typical of folds, faults, volcanoes, etc. are often found. If the rocks have been metamorphosed to roughly the same degree, the differences in resistance to erosion of major rock units tend to be insignificant; but smaller scale structural features, such as joints, foliation, or schistosity, may influence details of landscape development.

Whatever the structural type, when mountain elevations are high enough, low temperatures permit accumulation of snow that may turn to ice enabling glaciers to form, resulting in glacial landforms. In rare cases, where thick sequences of carbonate rocks (limestones, dolomites, marbles) are included in a mountain structure, large-scale complex erosional features formed by ground water may develop. The Dinaric Alps of Yugoslavia display spectacular landforms produced by solution action of ground water in these areas of complex geological structure (fig. 7.7, p. 134).

STREAM EROSION AND DEPOSITION IN DESERT REGIONS

Most of the explorer-geographers of the nineteenth century who made maps and wrote scientific reports about previously little known parts of the world were men who had grown up in humid temperate regions. The landforms, landscapes, and climatic conditions which they encountered in the extensive areas of sparse rainfall seemed strange and unfamiliar to them. Lacking both background and vocabulary to specify and adequately evaluate the natural phenomena they observed, explorers and surveyors tended to label as "desert" or "wasteland" any barren area without trees. The "Great American Desert," for example, was one of the most famous and persistent misnomers in geographic literature.

The basic criterion now used in classifying dry regions is mean annual rainfall. Any area receiving less than 5 inches of rain per year is considered *arid*; regions having between 5 and 20 inches of mean annual rainfall are *semiarid*. Although these parameters have reduced the areal extent of what used to be labeled "deserts" on the world's maps, the fact remains that nearly a third of the earth's land surfaces are dry enough to be described as arid (desert) or semiarid (near-desert).

What natural processes have produced the characteristic landforms and landscapes of the deserts? The impression brought back by explorers was that because deserts have hot, dry winds and piles of sand, wind action must be the dominant geologic process. This misconception has been corrected slowly. Strong winds do transport large amounts of earth materials in arid regions and in humid regions, too; but it is apparently true that in deserts, as in humid areas, running water is the single most important agent of erosion. (The wind's role as an agent of erosion and deposition is discussed in Chapter 7).

With a few exceptions, the processes of weathering and mass wasting, and the action of surface water and ground water all function in arid and semiarid regions about as they do in humid regions, but *at a much slower rate*. Fundamentally, most desert landscapes are the result of the processes discussed in this and previous chapters.

Because of the absence of moisture in the air, the rate of chemical and physical weathering in deserts is very low. The freezing that sometimes occurs at night and the high temperatures of the days do not have much effect on the processes of decomposition and disintegration of rock materials. Mass wasting under the influence of gravity, the shattering of rocks as they fall or roll down slopes, and the occasional but severe scouring occurring during runoff indicate that mechanical effects are probably more significant than chemical processes in the sculpturing of desert landforms (Cooke and Warren, 1973).

In the desert a light rain has little effect because of high rates of infiltration and evaporation. Characteristically, the total rainfall in a desert is scanty, spotty, and sporadic; and yet when rain comes, the brief period of runoff may be locally catastrophic. Since soil development is poor and vegetation sparse, loose rocks and sediments may be washed off desert surfaces in tremendous amounts.

Stream regimens are erratic, with occasional floods succeeded by months or years of dry channels. Well organized drainage patterns seldom develop. Because streams in desert country tend to flow with high volumes for short periods of time, many have large, anastomosing (braided) channels, choked with sand, gravel, and rocks. Other desert streams have eroded valleys with steep, vertical walls and wide, debris-choked floors. This type is called a *wash* or *arroyo* in western United States (Valyermo, California, fig. 3.5, p. 42), a *wadi* in North Africa. Sometimes runoff, concentrated back in the hills, surges down through an arroyo with high velocity and no warning, devastating whatever lies in its channel and dumping a new assortment of debris on the alluvial fans. Desert floods or seasonal runoff that comes down onto flatlands and dissipates through anastomosing channels spreads loose rubble around and then disappears—either from evaporation or infiltration. *Bahada* is the term for the extensive sheets of sand and gravel that may be built up in this way on the desert flats. Sometimes runoff collects in intermittent bodies of water called *playa lakes*. When the water evaporates or sinks into the basin, the flat expanse of sediment remaining is known as a *playa* (see p. 83).

Development of Pediments

Pediments are slightly sloping erosional surfaces—characteristic of arid and semiarid regions—that truncate bedrock and encircle mountainous units (fig. 4.13). As mountain slopes are removed, pediments grow larger. It is postulated that three activities in combination are responsible for pediment formation: (1) lateral planation by streams, (2) sheet and rill wash, and (3) weathering back of the mountain front. As the upland slope retreats, a sharp angle is maintained between it and the pediment surface. Theoretically, then, pedimentation is accomplished by the parallel retreat of slopes (fig. 3.1, p. 35). With the passage of time, as the mountains are worn back by slope retreat and pedimentation, only erosion remnants, called *inselbergs,* are left.

Travelers passing through desert country may have difficulty seeing the actual pediments below the mountain front because most of them are covered over with debris brought down by steep mountain streams and deposited in alluvial fans. Pediments truncating dipping beds (rock strata) can sometimes be seen in the sides of incising channels. The erosional surface at the top of the bedrock is the pediment.

Regional Drainage in Arid Regions

With runoff being discontinuous and insufficient, major stream systems seldom develop in desert environments. Also, because uplift of the land may have occurred, some of the world's arid regions do not have exterior drainage to the sea. Hence base level tends to be local only. In basins of deposition, such as the Wyoming Basin, which is surrounded by the ranges of the Middle Rockies, many thousands of feet of nonmarine sediments were laid down and subsequently rejuvenated by uplift so that terraces have been formed in dissected alluvium and in pedi-

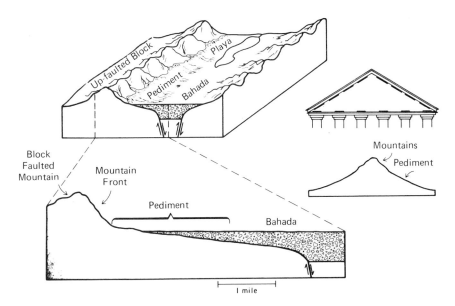

Figure 4.13. The cross-section on the lower left shows a pediment along the base of a mountain front, in this case an up-faulted block that has been eroded back several miles from the original fault scarp. The pediment is the erosional surface resulting from the retreat of the mountain front. Detritus eroded from the mountain and transported across the pediment collects in the down-faulted area and spreads over pediment, forming a depositional landform called a bahada. The diagram above the cross-section gives an overall view of the relationship of the up-faulted mountain block to the pediment and the bahada. Interior drainage is suggested by the presence of the playa (intermittent lake). The analogy of an architectural pediment to a geological pediment is illustrated on the right.

ments. In a few arid regions, rivers that originate in humid areas flow across the desert to the sea. Notable examples are the Nile, the Columbia, and the Colorado Rivers. In these drainage areas, ultimate base level exerts its influence, and the oscillation of basin filling and excavation is controlled by changes in river elevation. Pediment surfaces, bahada tops, and channels all are graded to the major streams.

Continental Sedimentary Deposits

In some regions of the world, deposits on a vast scale dominate the topography. Large quantities of rock debris (gravel, silt, etc.) have been eroded from upland areas, especially in semiarid environments, and deposited by sheetwash as blankets of sediments of great extent. East of the Rocky Mountains, continental deposits of Tertiary age (Ogallala formation) spread out over several thousand square miles in the Great

Plains country of Nebraska, Kansas, Oklahoma, and Texas. On this depositional landscape, modern streams are slightly incising themselves into the extensive sediments. In Argentina and Australia, streams have deposited immense composite alluvial fans and sheets of clastic material that are remarkable examples of depositional landscapes.

PUTTING THE PUZZLE TOGETHER

This chapter and those preceding it have discussed some of the dynamic forces of change—weathering, downslope movement, stream erosion, deposition, etc.—that continually act on the surface of the earth. The geologic work of running water is unstoppable; equilibrium is established and then disrupted as erosion, transportation, and deposition go on in a never-ending cycle. The processes of nature are influenced and regulated by many controls—climate, base level, structure, diastrophism, and so on. To study landforms and landscapes is to find all of these elements in complex interactions. In order to describe and explain the endless complexity, variation, and distribution of the earth's surface features, scientists bring forward terms to label them and suggest explanations, both speculative and pragmatic, for their existence. In many instances, the full truth is not yet known.

You, as a student, can pick out individual features and apply descriptive or genetic names to them. You can recognize on-going processes and formulate a variety of generalizations: "This process under these conditions produces this landform." During periods of rapid change, as in times of high water, you may be able to predict how an observed feature might be altered.

Look at photos and maps of the land; practice "windshield geology" while driving along highways. Recognize features, use your knowledge and understanding, and then come up with explanations for the origin and appearance of specific landforms. Take account of the complicating effects of multiple processes, discern the influence of different rock types, and then visualize and characterize by terminology and explanation areas too large to be seen at ground level. When you travel by jet, observe the drainage patterns on the surface of the earth below. Notice how structure has affected the landscapes and observe how they change as the jet speeds over a continent. Utilize your own observations and experiences as you get into the chapters that follow, which describe how others have suggested ways to rationalize and systematize the problems involved in attaining an understanding of landscapes.

SUMMARY

Streams, valleys, and drainage systems may be described and classified according to direction of flow, relief, texture, pattern, and degree of adjustment and integration. Laws of drainage composition have been developed from quantitative parameters of stream networks and basins. Processes operating in drainage systems tend to establish and maintain a condition of dynamic equilibrium or steady state. Climate, relief, and geologic structure are controlling and limiting factors in the development of fluvial systems. Tectonic activities and changes in base level may cause significant alterations in processes operating in a watershed. Most of the geologic processes operating in humid regions are also at work in arid regions, generally at a slower rate. Erosional landforms characteristic of arid regions are mountain slopes, pediments, inselbergs, arroyos, and dissected terraces. Depositional features are alluvial fans, bahadas, and playas. (Features formed mainly by wind are discussed in Chapter 7.)

Landscapes Produced Mainly by Stream Action

Examples illustrating importance of elevation and relief: Deep valleys and coarse-textured drainage in high mountains (northern Rockies in central Idaho); valleys in low mountains with moderately coarse-textured drainage (Tennessee, North and South Carolina, Georgia, Arkansas, western California); rolling hills with moderately fine drainage texture (eastern Ohio, central Missouri, southwestern Wisconsin); badland topography with very fine drainage texture (South Dakota, coastal Georgia, and South Carolina); canyon lands in areas of high elevation (northern Arizona, western Colorado, southern Utah).

Example of uplifting and rejuvenation: Valley in valley topography (northern Blue Mountains, Oregon).

Examples illustrating topographic changes over a period of time: Dissected upland flats with moderate elevations and low relief (Illinois and Iowa prairies); hill and dale topography all in slope with moderate elevation and relief (West Virginia, eastern Kentucky, central Pennsylvania).

Examples of topography formed mainly by alluviation: Floodplains, alluvial plains, braided stream topography, distributary stream topography, deltaic plains (Mississippi embayment, San Joaquin and Sacramento River valleys, Yukon River valley, parts of Gulf Coastal plain).

Examples of control by geologic structure: Plains (southern Alabama); plateaus (Colorado plateau); fold mountains—ridge and valley topography on a series of folds (central Pennsylvania, Maryland, and Virginia); domes and basins (Black Hills, South Dakota; Nashville Basin, Tennessee); block faulting—basin and range topography or horst and graben (Utah and Nevada; Teton Range, Wyoming; Sierra Nevada, California); volcanic mountains (Hawaii; Mt. Shasta, California; San Francisco Mountains, Arizona); complex mountains (Northern Rocky Mountains; the Alps; New Zealand Alps).

Chapter 5

Geomorphic Systems—Theory and Philosophy

TOPICS

The development of basic concepts in landform study:
  Origin of valleys, uniformitarianism, glacial theory,
  the Western surveys, the geographical cycle, peneplain
  concept, the arid cycle, Penckian slopes.
Dynamic equilibrium
Pediplanation
Physiographic provinces, regional geomorphology
Geographic geomorphology
Quantitative analysis and mathematical models

Processes that shape landforms and landscapes have been the substance of the preceding chapters. How might these commentaries and discussions of the controlling and limiting factors be fitted into an organized and systematic framework? Is it possible to develop a comprehensive, all-encompassing procedure to relate and explain the origin and evolution of landforms and landscapes? The seemingly endless variety of topographic forms and natural features making up the surface of the earth are not easily subdivided into groups, nor do they readily lend themselves to classification with keys. Genetic associations are often obscure. A further perplexity has been the question as to whether classification schemes should be descriptive only, principally quantitative (i.e., spatial distribution), or genetic. For about a century, students, teachers, researchers, and authors have sought to deal with these questions. In this chapter some of the attempts that have been made to perfect systems or schemes for the classification of landforms and landscapes are reviewed.

Unfortunately, the precise mechanisms by which landforms originate and evolve have not been and are not yet completely understood. Unlike many other investigators, a geomorphologist cannot design laboratory experiments to test adequately hypotheses which deal with, for example, continental land masses and eons of time. Therefore, the logical procedures of analysis (both mental and mechanical) are the principal tools available. In the philosophical sense, inductive and deductive reasoning are the foundation of all scientific speculation. *Deductive reasoning* leads or draws the mind away from a general principle—such as the major premise of a hypothesis—to one less general and more specific (which might be thought of as a minor premise) and combines them both in a conclusion. The conclusion is the way in which a general principle affects a more specific one. Induction is the mental opposite to deduction. *Inductive reasoning*, therefore, leads or draws the mind from specific facts toward general principles. An inductive hypothesis begins with a series of observations bearing on individual cases and instances and ends with a general principle applying to all of them.

In formulating geomorphic concepts, one procedure for the development of the correct or most likely generalization has been the *multiple working hypotheses* method. One begins by gathering all possible hypotheses that might bear on the problem. By careful analysis and testing (wherever possible), incorrect theories can be discarded, a process which should leave, eventually, one correct or most likely hypothesis. If all hypotheses can be proven untrue, then more must be developed and tested. Where evaluation of a particular hypothesis is impossible because of lack of information, this approach helps to define what types of data are needed. The actual harmonizing and evaluating of theoretical concepts is often difficult because of lack of sufficient factual data dealing with the processes at work on the earth.

THE DEVELOPMENT OF BASIC CONCEPTS IN LANDFORM STUDY

The Origin of Valleys

In the years just before and following 1800, the scientific works of a Scottish physician, James Hutton (1726-1797), and his associate, John Playfair (1748-1819), were published. These volumes became a major part of the scientific foundations of geology and geomorphology (Playfair, 1802). In a series of naturalistic observations about streams, Hutton and Playfair postulated that *valleys are formed by the streams that flow in them.* In other generalizations, based on field notes and reasoning, the two scientists emphasized the interdependence of many aspects of stream valleys and drainage systems. Their theories replaced indefinite and unformulated ideas implying that valleys were formed catastrophically or created in some mysterious or incomprehensible way and later on occupied by streams (Chorley, Dunn, and Beckinsale, 1964).

Uniformitarianism

In the first "modern" textbook of geology, Charles Lyell (1797-1875), an English geologist, set forth his explanation for the development of landforms within the philosophical framework of uniformitarianism, a principle based on the assumption that geologic events occurred in the past in about the same manner and at about the same rate as they do now. Lyell, who did a great deal of traveling, lecturing, and writing, taught that "the present is the key to the past." Uniformitarianism implies that by studying the nature of current natural processes, one can visualize how things happened in the past. Scientists now are not so sure that all geologic processes operate under an inviolate law of uniformitarianism. With this theory, how can one explain or accept periodic episodes of glacial climate that have occurred only a few times throughout geologic history? Or, how do we account for cycles of mountain-building that come at random throughout geologic time and proceed at seemingly erratic rates?

Louis Agassiz (1807-1873) and the Glacial Theory

Agassiz's magnificent induction that a great ice sheet had moved from the north polar regions down over northwestern and central Europe and had modified many topographic features, answered many questions that had puzzled early naturalists in Europe and North America. Agassiz's ideas were in no way a systematic approach to an understanding of landscapes, but he identified a fundamental process, *continental glaciation*, that prior to the 1840 publication of his theory, was unrecognized or disregarded as having been a significant modifier of the landforms and landscapes of large areas of the British Isles, Europe, and North America.

The Western Surveys (c. 1865-1900)

In the years of westward expansion following the American Civil War, the federal government supported explorations and surveys for the purposes of railroad location, mineral exploitation, and the gathering of information for topographic mapping and geologic interpretations. Major John Wesley Powell (1843-1902) led several expeditions in the region of the Colorado plateaus and the Grand Canyon. He recognized the significance of erosion and some of its controlling factors and perceived the importance of some of these relationships in the development of landforms. Clarence King (1842-1901), the first director of the U.S. Geological Survey, also led some of the early surveys in the West. Both by his own activities and through his influence on Congress in behalf of other geologists, King generated interest in scientific exploration and secured financial support for such projects.

The monumental contributions of G. K. Gilbert (1843-1918) form the core of modern physical geology. His penetrating analyses of the interrelationships between processes and landforms are the basis of classical geomorphology. C. E. Dutton (1841-1912) produced a voluminous study of erosional processes and landforms of the Colorado plateaus that emphasized the significance of major amounts of erosion.

During the last part of the nineteenth century, a large number of topographic and geologic maps were produced by relatively unknown but competent members of federal and state surveys. Besides making maps, these field geologists prepared voluminous descriptions of rocks and structures, wrote detailed reports regarding stratigraphy, economic geology, and ground water, and began the task of developing interpretations. Data from these published works, which were available to all, formed the raw materials used by the investigators and synthesizers who have followed.

DEVELOPMENT OF CLASSICAL GEOMORPHIC THEORY

Beginning with the publication of William Morris Davis' essay, "The Geographical Cycle," in 1899, a succession of conceptual ideas and theories have been proposed. The theme of Davis' fundamental contribution, *the geographical cycle* (later called by others the "geographic cycle," the "geomorphic cycle," or the "pluvial-fluvial cycle"), was his proposal that the sequential changes in landscapes wrought by time could be referred to in allegorical terms paralleling the life history of man. Davis worked out these analogies as he sought examples that he could use in teaching his students at Harvard. Probably influenced by Major Powell's articles on erosion by rivers in the American West, Davis suggested that a landscape of whatever history and configuration would undergo a sequence of changes until it was all eroded away (figs. 4.1, p. 47; 5.1, 5.2). The steps along this trail were called "stages," and were named "youth, maturity, and old age," with an "initial stage" added later. (See maps of Casey, Illinois, fig. 5.3; Draper, Virginia, fig. 5.4; and Hagood, North Carolina, fig. 5.5).

In the idealized example described in detail by Davis, he visualized that the initial landscape would have undergone a fairly rapid uplift from the sea, would have uniform seaward slopes, and be composed of homogeneous materials. When Davis was criticized many years later for making these conditions too simple and idealistic, he replied that this had been the simplest case he could imagine and that, of course, there could have been other more complicated situations!

Followers of Davisian concepts developed terminology for landscapes that evolved on uplifted old lands, coastal plains, volcanic materials, and so on. All, however, followed the basic ideas of uplift and then erosion during relatively long standstills of the earth's crust. If tectonic activity, such as uplift (fig. 4.6, p. 55), came along before the cycle was completed, *rejuvenation* was said to have taken place. In Davisian terms, the landscape became "youthful" and began again to progress toward old age.

Critics have pointed out that the word "cycle," as used by Davis, is misleading inasmuch as landscapes do not go around in a processional order from stage to stage, eventually returning to the beginning.

	Youth	Maturity	Old Age
Nature of the divides	Broad and flat	Narrow ridges or rounded summits	Low and discontinuous uplands (monadnocks); long distances from major rivers
Relief	Low but increasing	High (maximum)	Low and decreasing
Flatlands	On top of the divides	Region all in slope	Lowlands, peneplain
Efficiency of drainage	Poorly drained swamps and lakes on uplands well above base level	Drainage well integrated and adjusted; maximum number of streams	Poorly drained swamps and oxbow lakes on floodplains of larger streams

Figure 5.1.   The geographical cycle in a humid climate.

In nature, most cycles are interrupted frequently, some almost continually. It was also noted that if the rate of uplift did not equal the rate of erosion, the land could never get uplifted enough so that the cycle could begin. Davis agreed with the semantic criticism of the term cycle but felt that people understood what he meant anyway. As he conceived it, the cycle is primarily time dependent, with the time element being relative. If a region undergoing erosion is at a low elevation and underlain by rocks of low resistance, the absolute time necessary to reduce the area to the old age stage would not be long. But for a region of high elevation underlain by resistant rocks, reaching the stage of old age would require a much longer time.

Subsequently, Davis wrote other articles in which he recognized that climate might change during a cycle, with the result that different landforms might begin to develop. Using the phrase "climatic accidents," he visualized situations of aridity differing from a "normal" climate, as well as episodes of glacial activity. It is interesting to note that the climate Davis regarded as normal is that of northeastern United States and northwestern Europe; i.e., a humid temperate climate with precipitation evenly distributed throughout the year and wide variations between summer and winter temperatures. In making this assumption, Davis was no different from other scholars and travelers of his age, who accepted the notion that European (i.e., Western) culture provided a natural and logical standard against which the rest of the world could be measured.

During his long lifetime, Davis traveled a great deal, corresponded widely, and read virtually all the geologic literature published in English, French, and German. Relatively little was written about the geography and geology of the arid regions and the tropics until after Davis had published his major papers. Nevertheless, he was a keen observer. Had he lived or worked in the Transvaal, Western Australia, or Burma, perhaps his normal cycle of erosion might have been different.

The major factors that Davis believed control landscapes are (1) structure, (2) process, and (3) time or stage, with structure being the most important. His long experience in field geology had made him aware of the great variety of rocks and structures and the control they exert on the development of landforms and landscapes. This principle

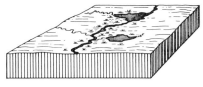

A. In the initial stage, relief is slight, drainage poor.

B. In early youth, stream valleys are narrow, uplands broad and flat.

C. In late youth, valley slopes predominate but some interstream uplands remain.

D. In maturity, the region consists of valley slopes and narrow divides.

E. In late maturity, relief is subdued, valley floors broad.

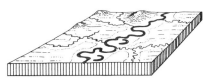

F. In old age, a peneplain with monadnocks is formed.

G. Uplift of the region brings on a rejuvenation, or second cycle of denudation, shown here to have reached early maturity.

Figure 5.2. Stages in the geographical cycle in a humid climate. (After E. Raisz in A. N. Strahler, 1969, **Introduction to Physical Geography,** 3rd ed. John Wiley and Sons. Copyright © 1969.)

has been articulated for generations of students as the Davisian equation: "Form = Structure + Process + Time."

Form, in this sense, means the shape of the landscape; structure includes the type and lithlogy of bedrock, geologic structure (fig. 4.11, p. 61), and mode of origin of depositional features; process deals with geologic agents; time or stage indicates the degree of progression of the cycle of erosion or the relative time that erosion has been working (fig. 4.1, p. 47). According to the Davisian equation, typical regions might be described as follows:

A folded mountain dissected by streams, in a mature stage.
A volcano glaciated by valley glaciers, in a youthful stage.
A glacial till plain eroded by streams, in a youthful stage.
A plateau eroded in an arid climate, in a youthful stage.

In all his writings, Davis sought ways of developing *clear, concise, genetic descriptions* of landscapes. Implied here is the idea of using genetically descriptive terms with especially coined or defined words to fit particular situations, rather than using general words from everyday speech. Precision of terminology would accomplish Davis' objective of replacing provincial and conventional terms for landforms by a modern scientific vocabulary and classification system. To give an example, the common words "hill" and "ridge" are simply descriptive; but "drumlin" and "hogback," used as landform labels, signify a certain precision of shape and a singular method of origin.

In recent years the Davisian cycle of erosion has been de-emphasized in teaching and in the literature as newer concepts and interpretations have come into use. However, the author, along with many other teachers of geomorphology, admits that no other approach thus far proposed to illustrate the evolution of landforms has equalled the Davisian cycle as a pedagogic device (Chorley, Beckinsale, and Dunn, 1973).

The Peneplain Concept

Davis postulated that at the end of the geographical cycle the land surface would be eroded essentially to base level. By this time, most of the uplands are worn away, and sluggish streams of extensive drainage

Figure 5.3. Casey, Illinois (1:62,500, C.I. 20 ft., 1942). The topography is typical of Davisian youth in a humid temperate climate because streams have been eroding here only since the retreat of the Wisconsin glacier. The dendritic drainage pattern is well developed, but streams have not cut deeply because the elevation is relatively low. Integration of drainage has not yet been accomplished. The criterion of adjustment does not apply inasmuch as the glacial drift (Chapter 6) is several hundred feet thick and essentially homogeneous.

systems wander over landscapes mantled with thick covers of alluvium. Drainage is not controlled by structure or differential resistance of rocks because with enough time, nearly all would have been removed. Old age is a theoretical stage, but if the geographical cycle ever proceeded that far, the resulting landform would be a *peneplain* ("pene" from Latin, meaning "almost"—hence, "almost a plain"). Peneplains would be subcontinental in extent and slope imperceptibly toward the sea. The term *monadnock* has been applied to any topographic highs remaining as uneroded hills, owing their existence to rock resistance or to position on a major divide. When Davis was asked, many years after he had proposed the idea of the peneplain, what the slope and relief would be, he replied that it was the kind of surface that "you could trot a horse on in any direction" (Cotton, 1958).

As a logical concept, the peneplain was a brilliant generalization; but ever since the idea was proposed, much discussion has gone on about whether peneplains ever existed, and if they did, whether they were formed in the manner suggested by Davis. At first Davis, followed by numerous imitators, began to recognize uplifted and dissected peneplains. This was done by drawing projected profiles across regions to determine whether accordance of summits existed (fig. 5.6). In many cases an accordance was present to a sufficient degree to allow the theoretical reconstruction of an earlier peneplain that had been subsequently uplifted (rejuvenated) and dissected by streams of the succeeding cycle of erosion. As time went on, enthusiastic "peneplainers" found accordance of summits on nearly all mountain ranges and on some, more than one level of accordance, the latter being interpreted as multiple peneplains. Standard theory has it that three (possibly four) uplifted and dissected peneplains (Schooley, Harrisburg, and Somerville) can be found in the central Appalachians and at least two in the southern and central Rocky Mountains (Flat Top and Rocky Mountain peneplains). Criticism has been directed at the validity of this method and such interpretations. Certainly too many uplifted and dissected peneplains have been described on too little evidence, but accordance in some degree does exist over large areas. For these, peneplanation may still be a valid explanation.

In general, criticism of the peneplain concept may be considered under two headings: (1) arguments against the peneplain concept itself

Figure 5.4. Draper Virginia (1:62,500, C.I. 20 ft., 1944). As can be seen from the contours, the region is all in slope, a characteristic of Davisian maturity in a humid temperate climate. The dendritic drainage pattern suggests homogeneous bedrock—either crystalline rocks or flat-lying sedimentary units. As the Draper map shows part of the Piedmont physiographic province—a region of metamorphic crystalline rocks—the former inference is the correct one.

and (2) suggestions of other ways in which peneplains might have been formed. Under the first category come arguments that tectonically the earth's crust is too active to remain at the same elevation long enough for a cycle of erosion to be completed. Elevation of base level would also be changing. At the present time, no peneplains are being formed that are generally agreed upon. Only the uplifted and dissected peneplains can be studied, and these are suspect. Erosion surfaces of low relief may be formed in other ways and confused with peneplains. Among such erosion surfaces are stripped or structural plains, pediplains (see page 67), plains of marine denudation, surfaces of lateral planation called "panplains," and exhumed former erosion surfaces.

In the light of such arguments, some geomorphologists feel that any application of the peneplain concept is invalid and misleading. Others feel that within the original assumptions of Davis, the peneplain hypothesis has a philosophical place in geomorphic thought; even though the concept may be of little practical use in the study of existing landscapes. Certainly the merit of the original hypothesis has suffered from too enthusiastic interpretations and from being given a degree of infallibility far beyond Davis' intention. The author believes that peneplains have been so much a part of regional geologic literature for so long that serious students need to understand how the concept has been used and how the map and field evidence has been interpreted. In regions where field evidence does show that flat-topped ridges and accordance of summits exist, any new theories will need to take account of such relationships (fig. 4.5, p. 54).

The Arid Cycle of Erosion

In W. M. Davis' geographical cycle devised for the drier parts of the world (1905), he suggested that, initially, elevation and relief were produced by tectonic or volcanic activity. In parts of the states of Nevada and Utah in western North America, for example, block-faulting took place during Tertiary time, producing mountain units with intervening basins—a topography and structure called "basin and range" (p. 63). Typically, ranges are a few miles wide and a few tens of miles long. The basins are roughly the same size. The uplifted blocks bordered by faults are termed *horsts* and the intervening basins are *grabens*. (Death

Figure 5.5. Hagood, North Carolina (1:62,500, C.I. 20 ft., 1938). Characteristics of late maturity or early old age in the Davisian cycle are evident here. The main stream has a wide, flat valley and an irregular course with swamps and oxbow lakes on both sides of the present channel. The valley walls are dissected and have gentle slopes (10 to 20 feet per mile). Because this region was not glaciated, the topography is most likely the result of long continued stream erosion.

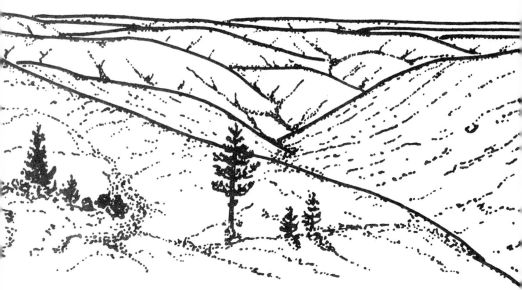

Figure 5.6. An example of accordance of summits across the Salmon River Mountains, Idaho. Bedrock here is granite of the Idaho batholith. Stream erosion has cut valleys over 3,000 feet deep into the uplifted area. (From Physiography of the United States by Charles B. Hunt, W. H. Freeman and Company. (Copyright © 1967.)

Valley and the Imperial Valley in California are very large grabens.) Elongated basins between mountains are called *bolsons*. In the western United States most of the areas of basin and range topography have developed typical arid region landscapes. Many have interior drainage with average annual rainfalls between 3 to 5 inches, with maximums up to 20 or 25.

The arid cycle begins with weathering and erosion attacking uplifted areas. In comparison with humid regions, the rates of weathering and erosion are slower and proceed more erratically. Because of the moisture deficiency, carbonate rocks are more resistant and tend to form steep slopes and rimrocks. Disintegration dominates the weathering process and produces quantities of rock fragments that tend to bury the bedrock outcroppings. When it rains, the hard, intense showers characteristic of arid regions sweep loose material from the slopes and dump the detritus along the mountain bases. Basin filling causes local base level to rise so that adjacent mountains can be worn down only to the level of the basin fill. In some cases, the basins fill to a point where detritus can be transported through a mountain pass and redeposited in a lower adjacent basin. Then the sediments from the higher basin proceed to fill the lower one (fig. 5.7).

	Youth	Maturity	Old Age
Relief	Maximum; initial deformation	Decreasing; degradation of mountains and aggradation of basins	Minimum; leveling without baseleveling
Drainage	Centripetal into basins; no trunks; erosion in highlands and deposition of alluvium in basins	Integration of drainage; low basins capture high basins; considerable transfer of sediments	Disintegration of drainage because of alluviation and wind action
Landforms	Basins of interior drainage and large areas of initial highs	Dissected mountains, aggraded basins, pediments, alluvial fans, bahadas, playas	Pediments completed and dissected; pediplains
Wind action	Some abrasion	Important local work by wind; dunes, blowouts, deflation	Dune building

Figure 5.7. The geomorphic cycle in an arid region.

Since the basin and range topography of the western United States is rare in other parts of the world, the details of Davis' cycle are of limited use. With different basic structures bringing about other types of initial topography in arid regions, any subsequent evolution of landforms results in patterns that vary from region to region. However, the sequence of the Davisian cycle is helpful in visualizing the development that is taking place since most geologic processes operate everywhere in similar ways.

Walther Penck (an enigma)

As an alternative to the erosional cycles of W. M. Davis, or to illustrate an adversary position, some students of landscapes have argued that the ideas of a German geologist, Walther Penck, could be used to refute Davis' theories about the evolution of landforms. Unfortunately, Penck died before his major work *Die Morphologische Analyse* (1924) was published. The German of his original manuscript has been described as "obtuse," and some confusion has developed as to his meaning. In translation the book is difficult to read. Though filled with original ideas, his writings cannot easily be reduced to a structured system that might be compared with Davis' work. Penck apparently suggested that the form of the earth's surface was the result of the interaction of endogenous forces (external or subaerial) and exogenous forces (internal or tectonic). Inasmuch as the earth's landforms can be studied and measured and the exogenous forces are known, one can use these data to better determine the nature of the internal or endogenous forces. In working this out, Penck classified slopes (valley side slopes) into three categories: (1) waning relief (decreasing) where the rate of stream downcutting producing concave slopes was slowing down; (2) waxing (increasing) relief when the rate of downcutting was increasing, producing convex slopes; and (3) uniform development where the rate of stream downcutting was constant, producing straight slopes (not concave or convex in profile). Penck suggested that by careful study of typical and average slope associations of a region, one could determine what tectonic activity had occurred or was occurring (von Engeln, 1940).

In a long paper published after Penck had died, W. M. Davis (1932), in making a partial rebuttal to Penck's ideas, inadvertently misread or mistranslated some of the complex and voluminous sections of Penck's book and other papers and attributed to him an advocacy of parallel retreat of slopes during erosional processes (fig. 3.1, p. 35). Since Davis' death, others have come forward to support Penck in his generalization (Bryan, 1940; L. C. King, 1953, 1962). Considerable opinion developed that the Penckian system of slope associations was a worthy successor or adversary to Davis' original concepts (von Engeln, 1942).

Not until 1953 was a complete English translation made of *Die Morphologische Analyse* (Penck; Czech and Boswell, 1953). Even this text is difficult to read and Penck's objectives are not clear. Recently Simon (1962) has carefully restudied Penck's works. He maintains that Penck never advocated the parallel retreat of slopes concept and that this was not a principal facet of his hypothesis. Simon says Davis misunderstood Penck's German. Simon further pointed out that most individuals writing about Penck used Davis' version of what he (Davis) thought Penck was trying to say. It would seem that Walther Penck's contribution lies in the area of slope study or careful analysis of the relationship between slope form and lithology, and the cause and effect relationships between valley deepening and gross slope form on the valley side walls. Penck did not advocate a system of geomorphology, nor did he consider himself an adversary of W. M. Davis. Stimulating as Penck's writings are regarding slope form and change with time, few geomorphologists have adopted his terms and ideas in their theoretical or practical studies and writings (Chorley, Beckinsale, and Dunn, 1973; Tuan, 1958).

DYNAMIC EQUILIBRIUM

Uplifted and dissected peneplains in the central Appalachian area have been described and argued about in geologic literature since before the beginning of this century. John Hack (1960) indicated that he was unable to accept Davis' premise of sequential changes of landscapes with time and thus joined other investigators who had looked at the flat-topped Appalachian ridges and been skeptical of the explanation that held them to be the result of multiple erosion cycles.

Drawing on the fundamental ideas of G. K. Gilbert (1877), Hack suggested that one could look at landscapes in terms of whether or not their various components are in a state of equilibrium. He postulated that when streams are ungraded—with waterfalls and flat reaches, and erosion taking place headward, downward, or laterally—the landscape is in a condition of disequilibrium. Slopes have differing angles, and the unequal resistance of various types of rock causes other irregularities in the landscapes.In a geomorphic sense, neither adjustment nor integration has been accomplished.

When the conditions prevailing are such that the drainage is for the most part integrated and adjusted, the major streams are graded. Various components, such as stream gradients, valley side slopes, floodplains, etc., are mutually interadjusted so that changes in any one segment bring about compensatory changes in other characteristics. Such a landscape is said to be in equilibrium. Conditions of equilibrium may change to disequilibrium because of orogeny, base level changes, climatic shifts, etc. The processes of erosion or deposition would then become shifted in the direction of the re-establishment of graded conditions and equilibrium. For those situations where Davisian theories or the peneplain concept are unsatisfactory, Hack's ideas and the dynamic equilibrium hypothesis are a welcome addition; but they are, nevertheless, difficult to apply. Data for individual valleys and streams can be collated, and determinations made as to graded or ungraded conditions; but correlating and interpreting data pertaining to the equilibrium of drainage systems and regions becomes quite difficult.

PEDIPLANATION

The key genetic aspect of Davis' concept of peneplanation is his hypothesis that with the passage of time, the slopes of valley sidewalls would retreat and grow flatter because of erosion. A somewhat different hypothesis for the reduction of uplands and the evolution of pediments was developed over a period of years by geologists who worked in the arid regions of the world. This theory takes the premise that once the slope angle has been established (through the interaction of climatic factors with a particular rock type), the upland slopes retreat in a parallel manner, always maintaining the same slope angle (fig. 3.1, p. 35).

After long experience in the continents of the Southern Hemisphere (in which are found large areas of arid and semiarid climates), L. C. King (1962) expanded this theory of the parallel retreat of slopes in an effort to make it applicable to other parts of the world. King disagrees with Davis as to the specific mechanism by which uplands are worn away and lowland flats produced. To King, the basic process is *pediplanation;* and the *pediplains* (the large areas of low relief characteristic of erosion cycles far advanced) are really complex systems of coalescing pediments.

Although superficially this might appear to be a matter of nomenclature, King's work represents a significant observation regarding the genesis of landscapes and a powerful criticism of the Davisian system. The crucial point here is the not yet completely solved problem of the retreat of slopes. The author can agree with King that pediplanation is a better explanation than peneplanation for landscapes in South Africa and central Australia; but whether pediplanation is the most satisfactory explanation for all fluvially produced landscapes is a problem that awaits more study.

ENVIRONMENTAL DYNAMISM

What is wrong with systematic theories of landscape development (i.e., those of Davis, Hack, L. C. King, Penck, and others) is the inescapable fact that most landscapes do not "fit" the theories. In some instances, the concept of uniformitarianism has been applied too literally or too narrowly in discussions intended to explain present-day topography. Davisian sequences and time-dependent explanations are possible, but probably did not occur in past environments exactly as they are seen to operate in the present. Basic physical laws do not appear to have changed. Events happening now, in accordance with these physical laws, also happened in the past; but some events happened in the past and were not repeated. There has been, in other words, an evolution of the earth's environments; and this problem, in relation to the principle of uniformitarianism, has never been fully resolved.

In a new synthesis of geomorphology, H. F. Garner (1974) emphasizes that the behaviors of geologic agents and processes must have been affected as the earth's matter evolved. Whatever atmosphere existed over the primitive earth, for example, is a subject for speculation; but the properties and effects of that combination of gases must surely have been different from the atmosphere we know now. The ancient sedimentary rocks that we see in some modern landscapes (e.g., Grand Canyon) have yielded all to few clues as to the actual appearance of the earth environment in which they were laid down; and the ancient landforms themselves have long since been destroyed. However, in a world of changing climates and widespread tectonic activity, the current processes of landscape development are controlled as much by the relicts of past environments as by the conditions of the present.

Environmental dynamism, as Garner views it, de-emphasizes landscape sequences and utilizes past and present combinations of geomorphic and tectonic processes. Earth dynamism is not wholly surficial (subaerial), nor is it wholly internal (tectonic), nor do processes and events appear to be precisely coordinated in a cyclical order. Previously proposed systems of geomorphology have not met realistic conditions of erosional and structural processes. Garner develops generalized explanations for the origins of landscapes that utilize past geologic history along with analogs from sedimentary environments and physical stratigraphy, changing climatic conditions, plate tectonics, and an acceptance that equilibrium or disequilibrium may occur irregularly during dynamic development.

PHYSIOGRAPHIC PROVINCES AND REGIONAL GEOMORPHOLOGY

While the systems of Davis and King approached the problem of landscape classification genetically and morphologically, another method of classifying landscapes using a regional orientation was being developed. Particular areas, it was found, tended to have distinctive landscapes and characteristic geologic features and could, moreover, be separated from other areas by natural boundaries in many instances. Such a definable land area is called a *physiographic province;* or, if more geology than geomorphology is to be included, the area might be termed a physical division or a physical province.

The landscapes within a physiographic province all have some similarities and significantly more similarities with each other than with landscapes of adjoining provinces. Ideally, boundaries between provinces are drawn along lines or zones of topographic change that are, more often than not, reflections of the underlying geologic structure. Rarely a physiographic boundary is drawn based on a geologic factor that may be indistinguishable at the surface. In the United States, the province boundaries were drawn by a committee of geologists (under the leadership of N. M. Fenneman), who prepared capsule descriptions of provinces and gave them vivid geographic and descriptive names, such as Glaciated Plains, Great Plains, Wyoming Basin, Reading Prong, and the like.

The usefulness of such a systematic organization lies in the fact that, knowing the general characteristics of a region, one can then predict the nature and type of landscapes and even the landforms. The degree to which this can be done depends on the distribution of varieties of landscapes and on the physiographer's skill in grouping like landscapes together and separating distinctive and significant features by province boundaries. When well developed, the system demonstrates the relationships of landscapes to the underlying rocks and their geologic history. In studies of small areas, however, where detailed geologic and topographic data are available, the delineation of meaningful regions separated by obvious boundaries may be more difficult.

Fenneman's outstanding works, *Regional Physiography of Western United States* (1931) and the companion volume, *Regional Physiography of Eastern United States* (1938) are valuable reference texts. Others in this field who have made contributions are A. K. Lobeck, who prepared a great many physiographic maps of portions of the world's surface; and Bruce Heezen and Marie Tharp (1958, 1962, 1964), who have depicted the landscapes of the ocean floors.

Publications on regional geomorphology have brought in geologic history, especially that of the last few million years, in order to explain the present nature and configuration of the earth's surface. Both W. D. Thornbury (1965) and C. B. Hunt (1967) have done outstanding work on the geological geomorphology of the United States. They have dealt with recent geologic history and drawn extensively on data about rocks, structures, and tectonics to illuminate more clearly present-day landscapes.

GEOGRAPHIC GEOMORPHOLOGY OR LANDFORM GEOGRAPHY (FROM 1950 ONWARD)

Within the allied disciplines of geology and geography, the scope and emphasis of geomorphic studies have been matters of considerable concern and debate over a period of years. Traditionally, geomorphologists, having been trained in geology, used geologic methods to develop explanations for the origins of landforms and landscapes. Their expositions were genetic in emphasis and cloaked in descriptive terminology. They drew on historical geology for explanatory concepts and used present-day landforms to develop further details about the more recent parts of the geologic history of a region.

Modern geographers, dissatisfied with the classical approach and its emphasis on the history and evolutionary development of landscapes, became more interested in landforms as a base for human activities. Some geographers began to develop for terrain analysis a descriptive terminology that was more appropriate for such purposes. J. E. Kesseli (1954) and other geographers advocated that investigators devise methods for quantitative empirical descriptions of landform type regions. After being trained in the recognition and definition of landform types, geographers could then map the landforms themselves by using appropriate symbols, in this way constructing morphographic maps. E. H. Hammond (1954), also a geographer, proposed using an empirical descriptive analysis, followed by an explanation for the origin of the landform, together with an analysis of the landscape in relation to other geographic phenomena—economic, cultural, physical, etc. Arthur Strahler (1954), as a geological geomorphologist, suggested that beginning students in geography should use explanatory, descriptive methods in landform study and that advanced students and researchers should use empirical-quantitative methods. Other geographers took the position that landform geographers should use the historical procedures of the geological geomorphologist and the descriptive quantitative methods of the geographical geomorphologist, pointing out that a study of geologic processes and historical geology is as important in landform geography as terrain description and analysis.

Implied in these discussions are criticisms of a too ready acceptance of Davisian genetic descriptions, where based on too few field observations, and a questioning of a general attitude of subjectivity in the preparation of landform descriptions. What is needed, say the geographers, are reliable quantitative data collected by proper sampling methods from accurate, up-to-date maps. The geographers recommend that large-scale maps be used (none smaller than 1:24,000, with a contour interval of 5 or 10 feet). Only by objective collection of unbiased quantitative data, say the purists, can valid landscape descriptions be developed. Once data are available, they can be organized and studied in terms of any of the following aspects: Slope analysis, local relief analysis, texture and roughness analysis, pattern and profile analysis, regional synthesis, functional landform analysis, landforms and other physical phenomena, landforms and cultural phenomena, covariation of landform elements, and spatial variation and distribution of landforms (adapted from Zakrzewska, 1967).

While it would be possible for masses of landform data to be collected objectively, stored, and analyzed with the aid of computers so that theoretical principles of landform geography might be worked out, thus providing a solid basis for meaningful studies, the practical difficulties are great. Many maps are insufficiently accurate, and not enough large-scale maps are available, even in the United States. Furthermore, data-collecting from maps, which must be done by highly skilled persons, is exceedingly tedious and time consuming. And yet, unless data are collected in the same manner from basically similar maps, the quantitative comparison and integration of the results are relatively valueless. However, from such data as are available and relevant, the formulation of theoretical principles can be continued. A number of geographic geomorphologists in different parts of the world are already working along these lines. Satellite photography, by which remarkably good detail over large areas can be produced, is being used, for example, to map the growth of Arctic and Antarctic pack ice and drought conditions in the southern Sahara. This type of photography and other imagery can be obtained by remote sensing techniques with much greater speed and at only a fraction of the cost of conventional topographic maps—and may eventually replace them (pp. 8-9).

A World-wide System of Landform Classification

R. E. Murphy (1967, 1968) has proposed a world-wide classification of landforms that is both genetic and descriptive and utilizes empirical components as well. Murphy has developed three levels of categories that may be outlined as follows: (1) seven structural regions—alpine system, Caledonian remnants, Gondwana shields, Laurasian shields, rifted shield areas, sedimentary covers, and isolated volcanic areas; (2) six types of topographic regions—plains, hills and low tablelands, high tablelands, mountains, widely spaced mountains, and depressions; (3) five classes of areas dependent on geomorphic process—humid landform areas, dry landform areas, glaciated areas, Wisconsin and Würm glaciated areas, and icecaps.

Thus, for example, the Gulf Coast of the southern United States can be classiifed as *sedimentary cover* (structural region), *plains* (topographic region), and *humid landform* area (geomorphic process). The Canadian Rockies of western Canada may be designated as *alpine system, mountains,* and *Wisconsin glaciated area.* This classification system uses the genetic characteristics (i.e., structure, process, and time) in the first and third category. Quantitative parameters, such as elevation and relief, are utilized in the second category. While not enough detailed geologic and topographical data are available as yet for some areas of the earth's surface, this type of classification does represent a very good basic attempt to encompass all of the world's topography.

QUANTITATIVE ANALYSIS

Robert Horton's (1945) contributions to scientific study of drainage basins (p. 48) laid out a framework for the quantification of geomorphology. Over a period of time Arthur Strahler (1964) modified, extended, and developed Horton's ideas, making them applicable to a broader range of surficial phenomena. Strahler advocated the use of statistical procedures for handling the voluminous data collected from maps and from the field in order to attain a sounder basis for both practical and theoretical work than could be gotten through the "arm-waving and eyeballing" methods of earlier investigators. For example, in a 1950 paper ("Equilibrium theory of erosional slopes approached by frequency distribution analysis"), Strahler determined that a comparison of valley-wall slopes with adjacent channel gradients reveals a strong positive correlation, indicating a high degree of adjustment among component parts of a drainage system (see p. 49). In a new conceptual approach, Strahler (1952) proposed using a series of parameters (gravitational stresses, molecular stresses, chemical processes, etc.) as a *dynamic basis of geomorphology.*

In retrospect, morphometric analysis, which employs quantitative descriptive values, enables an investigator to proceed along three major lines: (1) to describe landforms using numerical values; (2) to express comparisons quantitatively, i.e., twice as large, 60 percent greater, etc.; and (3) to utilize numbers in analytical and statistical procedures in order to develop generalizations, relationships (such as ratios), and "laws." By such means, humanistic terminology (*e.g.,* "treeless wastes") and the genetic terms of Davis and others are avoided. The procedures of quantitative analysis can be more objective and also precisely descriptive without implying process, sequence, or passage of time. Statistical methods may be used to discover relationships too complex to be seen "by eye," and comparisons can be made objectively (correlation).

Data can be tested for internal reliability before being used statistically and may be analyzed in many ways—such as by asking: As this factor changes, what are the relationships to changes of another factor? Most significantly, the data can be processed by electronic computers.

In programming a computer, an investigator needs to make careful judgments and selection regarding data and methods. As E. F. Morison has said in *Men, Machines, and Modern Times* (1966), computers can do many things: "remember, learn, discern patterns in loose data, make novel combinations of old data, and, most striking of all, introduce surprise into an intellectual situation . . . [but] if we put the wrong things into it, if we select the wrong problems or state the right problems incorrectly, we will get unsatisfactory solutions. . . . In using the computer, man will get the answers he deserves to get."

With these cautionary aspects duly in mind, it must be emphasized that the scientific value of studies using quantitative analysis is that these approaches can greatly widen and sharpen our understanding of geomorphic processes and the resulting landforms. With quantitative relationships evident between various aspects of a terrain and its streams, it follows that definite, orderly relationships exist also between processes and landscapes. That landforms can also evolve proportionately to each other seems probable. The true significance of these numerical relationships is that insofar as they show processes producing landforms and landscapes in an orderly manner, it must be assumed that the processes themselves are in some way regulated and controlled. Further study of the dynamics of geomorphology may delineate better the reasons for these fundamental controls. Some advanced research now underway is heading in this direction.

MATHEMATICAL MODELS

A geomorphologist may formulate—through invention or intuition based upon the sum total of his experience—a mathematical model that is a quantitative statement of theory, a situational relationship, a process, etc., previously definable only in descriptive or qualitative terms. A few such models have been developed for wave velocities, landslides, dune migration, etc. Digital computers are used to simulate some of the complex dynamic systems in geology and geomorphology. Examples are models of sedimentary cycling (of minerals and compounds), a sedimentary basin model, a delta model, and a carbonate ecology model (Harbaugh and Bonham-Carter, 1970).

Many of the conclusions in geology and geomorphology are based on inductive reasoning; one starts with observations of the landform— the end product— and reasons backward in order to determine the method by which it was formed. Laboratory and scale model experiments are useful in enlarging our understanding of current geomorphic processes at work on the earth's surface and can serve as a guide in investigations of how present landforms have evolved. Because of the severe handicap of researchers' being unable to scale down sufficiently materials, time, and geologic agents, scale model experiments cannot always yield true quantitative relationships but can show causes and effects, as well as other interactions. Large flumes and basins—such as those at the Coastal Engineering Research Laboratory near Washington, D.C., and at the Mississippi Waterways Experiment Station at Vicksburg, Mississippi, in which stream channels, beaches, harbors, etc., can be built in scale size—have proved to be useful in the designing and planning of many engineering projects.

SUMMARY

The publication of W. M. Davis' "The Geographical Cycle" in 1899 marked the beginning of the systematic study of landscapes, with genetically descriptive terminology and the use mainly of deductive reasoning, based on his concepts of structure, process, and stage (time). Conceptual schemes were later developed involving grade, equilibrium, and mathematical models. Robert E. Horton, Arthur N. Strahler, and others, introduced procedures for quantitative geomorphology based on measurements from maps and other statistical parameters. This inductive approach to the understanding of geomorphic processes began with the study of stream flow data and channel and valley morphology (see pp. 48-51). Evaluation of the role of climate in the control of processes and rates has also been an important trend. Regional geomorphology, as developed by N. M. Fenneman, A. K. Lobeck, W. D. Thornbury, and others, has shown how relationships of geologic and climatic factors have tended to produce similar or dissimilar physiographic provinces or physical divisions. Geographical geomorphologists have proposed a spatial framework for landscape study to supplement genetic and descriptive systems (Dury, 1969).

Chapter 6

Glaciers and
Glaciated Landscapes

In their beauty, remoteness, and association with challenge and danger, glaciers have long had a fascination for explorers, mountain climbers, and skiers. Not surprisingly, inquisitive persons have long been attracted to study of glacial phenomena. Since Agassiz's proposal and development of the "Ice Age" theory during the middle of the nineteenth century, a large body of scientific data has been assembled indicating that widespread glaciation in the recent geologic past has been the geomorphic agent responsible for many characteristics of the world's landscapes.

CLASSIFICATION OF GLACIERS

Glaciers are found in a great variety of sizes and forms around the world, ranging from cirquelet glaciers a few hundred feet wide in the high Sierras to the Antarctic icecap, the largest existing glacier. By means of morphological criteria, glaciers can be grouped into four general types: (1) *Cirque glaciers,* located at the heads of valleys in mountainous regions, are small masses of ice, shaped roughly like half moons, and are, in most cases, remnants of much larger glaciers. (2) *Alpine* or *valley glaciers* are long, narrow rivers of ice that originate high in mountains and descend through valleys to lower elevations (Mount Rainier National Park, fig. 6.1). Many are 5 to 10 miles in length, but a few extend as much as 100 miles. (3) When several valley glaciers debouch from a mountainous area and join together in a wide, bulbous ice mass, they form a *piedmont glacier*. (4) In some climatic environments, so much ice accumulates that the underlying topography is completely buried under an icecap or *continental glacier*, several miles thick and of immense dimensions.

Good examples of cirque glaciers can be seen in the Sierra-Cascade range of western North America. Alpine glaciers in Switzerland and New Zealand have been photographed and mapped and are accessible to travelers. Relatively few piedmont glaciers exist, in comparison with other types, but good examples are located in southeastern Alaska in the Chugach mountains, in the St. Elias mountains on the Yukon border, and along the foot of the southern Chilean Andes. Active continental glaciers, in addition to the Antarctic icecap, are the Greenland icecap and the glaciers of Baffin Island, Iceland, and Novaya Zemlya.

A thermal or geophysical classification of glaciers recognizes *temperate glaciers* that have temperatures corresponding to the melting point of ice; *high polar glaciers*, where temperatures the year round are below the freezing point; and *subpolar glaciers*, where the surface temperature gets up to the melting point for short periods of time. Temperate glaciers produce copious runoff, but polar types have little or none. Glaciers can also be classified dynamically as to whether they are *active* or *passive*. In general, active glaciers occur in maritime environments at relatively low latitudes; passive glaciers are found in high latitudes in continental environments (Cotton, 1947; Flint, 1971).

GLACIAL ICE

In regions where more snow falls than melts each year, the excess, as it accumulates, keeps piling up on top and packing down and recrys-

Figure 6.1. Mount Rainier National Park, Washington (1:62,500, C.I. 100 ft., 1955). The valley glaciers eroding this stratovolcano (Chapter 7) were much more extensive during the Pleistocene. Then as now, moisture-laden winds from the Pacific dropped quantities of snow on the windward side of the mountains. Snowfall at about the 10,000-foot elevation on Mount Rainier is estimated to be the equivalent of an average of 100 inches of rain per year. Willis Wall on the north side of the summit and Sunset Amphitheater on the west are cirque headwalls. The "cleavers" (a local term) are arêtes. The dotted line on the surface of Ingraham Glacier represents a medial moraine. The speckling on the snouts of glaciers indicates the presence of ablation moraines. The Nisqually Glacier (p. 97) is south of the peak.

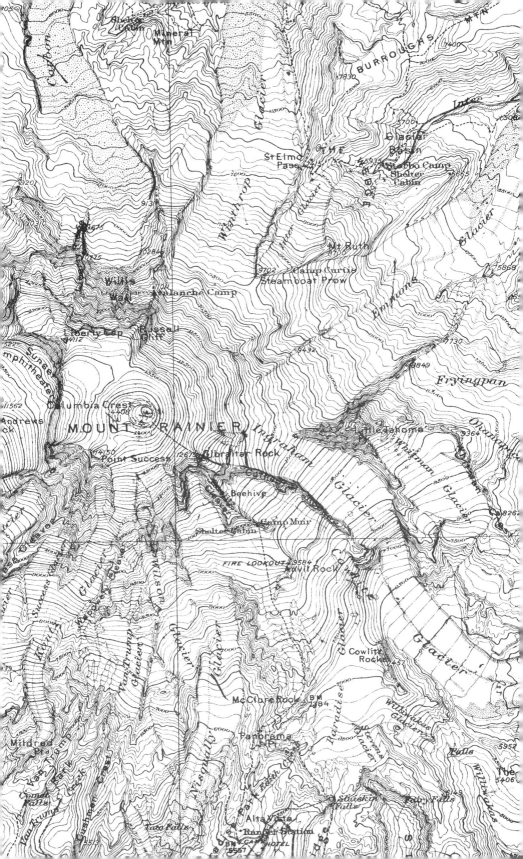

tallizing the snow underneath. As the mass of snow and ice gets tighter and tighter, air is pushed out. Rain and intermittent melting of surface snow produce water that drains down into voids and cracks. Subsequent freezing tends to compact the entire mass into dense, solid ice. On a small scale, schoolboys in cold climates have long known the method. They scoop up double handfuls of soft snow and press it together, shaping it into hard balls. When these are soaked in water and then refrozen, they become missiles of solid ice!

Since ice and snow cannot withstand much stress, the increasing pressure of the accumulating mass plus the force of gravity causes the lower ice to move downward or outward, thus initiating glacier movement. Glaciers tend to move from their areas of accumulation toward areas of warmer temperatures where melting or wasting exceeds inflow. In the zone of accumulation lie the *perennial snowfields* that nourish the glaciers. Below is the zone of wastage. Separating the two zones is an irregular shifting band called the *snowline,* or firnline, that determines the elevation of glaciers. In the tropical regions of the world, the snowline is more than 15,000 feet high, but it descends to sea level in the Arctic and Antarctic. Between global high and low points, the snowline dips lower in areas of high rainfall or cool summers and climbs higher in dry, hot regions. In detailed mapping, the snowline is higher on warm, south-facing slopes and lower on cooler, north-facing slopes. (The reverse is true in the Southern Hemisphere.) The actual position of the snowline varies from season to season. Long-range fluctuations are caused by slowly shifting, world-wide climatic changes. Modern glaciers are found in regions of high altitude and/or high latitude where sufficient precipitation is usual (*e.g.*, a high mountain range along a rainy coast) (Dyson, 1962).

GLACIER MOTION

Measurements of glacier movement indicate that velocities range from fractions of an inch to several feet per day. In the case of valley glaciers, the velocity changes from side to middle and from bottom to top, the faster movements being near the middle of a glacier at its surface. This pattern is similar to that of streams and is the result of friction between the ice and the confining valley walls and floor. Ice in continental glaciers moves more slowly except in places where narrow tongues of ice are forced out through mountain passes, as in the case of the outlet glaciers of the Greenland icecap.

The mechanics of glacier motion are not completely understood; but a number of field, laboratory, and theoretical studies have indicated that movement is accomplished by (1) external stresses acting on the overall mass of ice and by (2) internal stresses on the crystalline structure of the ice as it responds to external force. With the first mechanism, the weight of the ice itself and the force of gravity cause the glacier to slide over the rock walls and floor of its valley. This motion, called *slip*, might be described as a slow type of avalanching. Slip is probably the chief means by which thin ice moves down steep slopes. Since ice is a brittle solid, *shearing* and fracturing, which allow masses of ice to move over or around other parts of a glacier, also contribute to glacier motion. Crevasses, formed by shearing or cracking in the surface zones of glaciers, provide some evidence of this mechanism. However, crevasses seldom extend more than a hundred feet into the ice, and moving glaciers are usually much thicker than that. It would appear that slipping, shearing crevassing, etc., which operate mainly underneath, on top, and at the sides of glaciers, could not account for the massive and complex overall movement of great bodies of solid ice.

An explanation for the second type of glacier motion, called *plastic flowage,* can be found in the inherent characteristics of ice itself. Because ice can be deformed and made to "flow" rather easily under stress (pressure), it is believed that the main mechanism of glacier motion involves complex slipping or sliding between crystalline grains of ice (intergranular shifting) and along glide planes within the ice crystals (intragranular shifting), accompanied by continuous recrystallization. Inasmuch as ice crystallizes in the hexagonal crystal system, stress applied in one direction may result in crystal gliding, but applied from another direction may bend, deform, or crush the crystals. When ice grains are pressed together under pressure, they may melt and slip. As soon as the pressure is relieved, the ice refreezes. Pressure, increasing with depth of the ice, increases the amount of plastic flow, as well as the melting and refreezing of ice grains (Sharp, 1960).

Because glacier valleys are irregular in shape, the pressure and flow conditions vary considerably—as when glaciers move around bends or pass over "ice falls." Some measurements made at the surface of glaciers show irregular and jerky motion, suggesting that glacier flow is being accomplished by several methods at the same time. Moreover, since there is usually a gain or loss of ice along the surface of glaciers as they move downward, velocity tends to increase ("extending flow") or decrease ("compressing flow") accordingly.

GLACIAL REGIMENS AND CLIMATIC FLUCTUATIONS

Changes in the amounts of ice flowing in glaciers and the continual advances and retreats going on at the snouts and margins reflect variations in precipitation and temperature that may be seasonal (short range) or an indication of longer trends. Brief variations in the velocity of valley glaciers appear to be related to daily and seasonal fluctuations of the air temperatures. When a sequence of years of heavy snowfall enlarges the snowfield, the increased accumulation creates a "bulge" of ice that works its way downward more rapidly than the average velocity. Such a "kinetic wave" on the Nisqually Glacier (fig. 6.1) on Mt. Rainier,

Washington, traveled down at a speed of about 700 feet per year (during the 1940s and 50s), while the average ice velocity was about 200 feet per year (Johnson, 1960). The crest of the wave traveled faster than the trough succeeding it, causing a change in the shape of the glacier. When a thickening wave of this type reaches the brittle ice of the glacier snout, it may ride up over the lower ice in thrust sheets (Meier, 1964).

Within historic time, periods colder than the present have been recorded. The most recent occurred from about the sixteenth century through the eighteenth. Mountain passes in the Alps used by Hannibal on his way to invade Rome (218 B.C.) were closed during the late Middle Ages and are now open again. Climatic shifts of much longer duration are believed to have controlled the advances, standstills, and retreats of the margins of continental glaciers of the Pleistocene Epoch. When glaciation was at maximum extent, snowlines were as much as 4,000 feet lower, and glaciers existed on nearly all the mountain ranges of the world, besides covering the northern regions of Europe, Asia, and North America. The effects of glacial ice and meltwater can still be seen in landscapes all over the world.

In the last three decades, mean global temperatures have dropped slightly and Arctic ice has increased in the last few years. Some mountain glaciers have advanced in scattered locations. Whether these advances are due to local conditions or to slight climatic shifts has not yet been determined. Some climatologists and environmentalists have suggested that man's activities involving material discharged into the atmosphere, among other factors, may be causing perceptible climatic changes and even an eventual return to large-scale glaciation within the next several hundred years.

METHODS OF GLACIAL EROSION

Eroding glaciers do much of their work by *abrasion*. Loose pieces of rock of all sizes and chips or grains of rock frozen in the moving ice serve as grinding and gouging tools. On glaciated bedrock, the presence of polished, scratched, chipped, gouged, and grooved surfaces testifies to the severity of glacial abrasion. Glacial scratches or grooves, called *striations*, are one of the essential indicators of past glaciation.

As a glacier overrides obstructions, it erodes by *plucking* or *quarrying*. The weight of the ice impinging on the edge of a rock mass can crack, chip, or flake off large pieces of an outcrop. Ice freezes onto masses of cracked or shattered bedrock; then, as the glacier moves on, it plucks or quarries out blocks of rock. Some bedrock surfaces, especially on lee slopes, are jagged and irregular where loose blocks, large and small, have been torn out. Presumably, boulders identified as glacial *erratics* are a result of these forces, the size of the erratics giving some indication of the scale of crushing and plucking. Some erratics were carried in the ice many miles before being dropped. (Erratics are not true landforms—although a few are large enough to qualify.)

Alpine glaciers erode their valleys headward, forming steep-walled cirques, which resemble bowls gouged into the rock and tilted on edge. The glacial ice looks as though it were flowing out over the rim, like thick batter sliding off the lip of a mixing bowl. The method of cirque formation is still somewhat uncertain. Alpinists have long known of the presence of the *bergschrund* (fig. 6.2), a large, semicircular crevasse that separates the back of the glacier from the cirque headwall and forms an opening between the upper slope of the glacier and the snowfield above. Down-valley movement of the glacier keeps opening the bergschrund, and snowslides, avalanches, and meltwater continually fall or cascade into it. Intrepid climbers who have made descents into bergschrunds have told of finding perpendicular bedrock on the uphill side and a wall of ice on the downhill side. Both walls were dripping wet and showed evidence of extensive freezing and thawing. The headwall side was cracked and riven by frost action and plucking, and the ice wall was studded with broken rock ready to be carried away.

Since air circulates in a bergschrund and plenty of water is available during times of thaw, the rate of erosion is high. Under these conditions, it seems plausible that a glacier might "eat" its way headward and slightly deepen its cirque. However, glaciological studies have suggested—on the basis of data taken from automatic recording instruments—that the freezing-thawing-plucking hypothesis for cirque formation may be oversimplified. Apparently frost-shattering and plucking by the ice takes place mostly at the top of the bergschrund, where the temperature

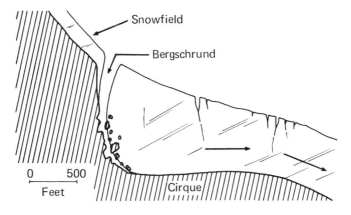

Figure 6.2. Cross-section of a bergschrund at the head of an alpine glacier, showing plucking of the headwall, abrasion and deepening of the cirque.

fluctuations are most extreme. Farther down in the crevasse, cascading meltwater washes down the loosened material with great force, eroding and deepening the cirque and, presumably, giving it the characteristic "down-at-the-heel" configuration.

Most of the processes of glacial erosion discussed above assume a rigid or brittle condition of ice, such as that found on the upper part and sides of the glacier. And yet the deeper part of a glacier is regarded as flowing plastically under pressure. Thus it would appear that the upper parts and edges of the glaciers would necessarily do the bulk of the eroding. However, the nature of glacial topography and the vast amounts of debris which have been transported seem to indicate that ice throughout the glacier has to be capable of accomplishing considerable erosion and transportation. Since the practical difficulties involved in the collection of empirical data from the critical parts of glaciers are very great, it may be some time before completely satisfactory explanations for some of the complex problems of glacier movement and erosion are developed.

Glacial Transportation

Continental and valley glaciers transport debris on top of, within, and frozen to the bottom of the ice—in other words, superglacial, englacial, and subglacial. Because glacial ice is highly viscous, it can carry a much larger sediment load than water. So much debris collects in the zones of wastage that it is not unusual for glacial ice to be completely covered with detritus (an *ablation moraine*), dozens of feet thick. Trees ride along on debris blanketing the Malaspina glacier, a large piedmont glacier in Alaska. Talus material, broken off side walls above the edges of valley glaciers, falls or rolls down and accumulates as ridges of debris along the ice margins. These *lateral moraines* are carried along the edges of a glacier as it moves downward. When valley glaciers begin to thaw, the debris acts as an insulating blanket, preventing rapid melting and sublimation of the ice. Some cores of ice have lasted for hundreds of years buried in moraines near glacial snouts.

Glacial Deposition

Glacial drift is an all-inclusive term for any kind of rock material—boulders, gravel, sand, clay, etc.—transported and deposited by the ice or meltwater of glaciers. It is difficult to visualize the tremendous quantities of drift that have been spread or dumped or laid down over the continents and the adjacent sea floors. Drift blankets some areas and exists in discontinuous patches or spots in others. In general, the contrast between regions that have undergone glaciation and hence have drift deposits and those that have none is noticeable enough so that it can be detected by even a casual traveler.

Unsorted and unstratified deposits that were put down underneath the glacial ice or dumped along the edges are called *till* by geologists. Comprised of material of many sizes and kinds jumbled together, till deposits may be loose but are more often found tightly compacted. Till deposits are known as "boulder clay" in the British Isles and "hardpan" in the eastern and central United States, both terms being descriptive. The word "moraine" is sometimes used interchangeably with till—although, strictly speaking, moraine refers to a specific group of ice-deposited topographic features composed largely of till.

Stratified drift is a general term used to designate glacial drift composed of water-laid material of glacial origin. Related types of deposits might include glacial lake deposits, glacio-marine sediments, and wind-transported silts (loess) blown off the surfaces of glaciers or from the tops of glacial deposits. The stratified drift dropped by meltwater accumulates both within the margins of a glacier—these sediments being called "inwash"—and beyond the edge of the ice, in which case they are termed "outwash." At the lower ends of valley glaciers and over large areas of ice sheets, meltwater flows over the ice surface, spills down into crevasses, bursts out again from tunnels under the ice, and collects in hollows on the ice or in pools damned by ice. Running water is everywhere, much of it from melting ice, but a good share coming from rain and runoff that has drained down onto the glacier from nearby uplands. With quantities of debris of all sizes available, the running water soon has a sediment load to transport—which it subsequently deposits in a great variety of forms.

Glacial Landforms

The glaciers of the world have been so varied in their attributes that their erosional and depositional processes have produced a multiplicity of landforms (Embleton and King, 1968). For purposes of convenience and comparison, typical features have been arranged in tabular form (fig. 6.3) together with descriptive information about their shapes, compositions, and origins. Some of the landforms characteristic of valley glaciers have names of French or German derivation because it was in the Alps that the first scientific studies of glaciers were made.

The Nature of Deglaciation

Evaluation of information about the size and distribution of landforms in glaciated topography can illuminate the nature of past ice movement and subsequent deglaciation. For example, for a glacier to build a large end moraine, its edge would have to remain in a location for several hundred years, with the amount of melting roughly equaling the amount of new ice pushing up to the edge and dropping its debris. Stagnant glaciers (that are not bringing down fresh supplies of ice and dirt) cannot build end moraines. When thawing exceeds the rate at which ice is brought to the edge, the glacier begins to recede, and its load is spread over the surface as ground moraine. The presence of meltwater deposits of various types and sizes suggests that ice stopped moving and became stagnant, inasmuch as moving ice would have destroyed any

I Erosional Features	Topographic Description	How Formed
Horn, matterhorn (vg) *	3- or 4-sided, pyramidal peak standing on a divide.	High altitude weathering and headward erosion of coalescing cirques.
Col (vg)	Crescent-shaped pass between horns on alpine skyline.	High altitude weathering and headward erosion by cirques back to back.
Arête (vg)	"Knife"-edged ridge on sides of horns, or between cirques or troughs.	High altitude weathering, glacial plucking, and abrasion.
Cirque, corrie, cwm (vg) (Rock-dammed lakes in cirques are *tarns*.)	Steep-walled, semicircular valley head.	Plucking, abrasion, nivation.
Glacial trough, U-shaped valley (vg) (*Fiords* are "drowned" glacial valleys.)	Steep-sided, flat-floored, open valley.	Preglacial stream valley modified by abrading and plucking.
Glacial staircase (vg) Rock-dammed lakes on flat segments of staircase are *paternosters*.)	Alternating steep and flat segments on long profile of glacial valley.	Plucking and abrasion; location of steps, steepened by quarrying and crushing, probably determined by concentration of bedrock joints.
Hanging valley (vg)	Tributary valley with floor much higher than the elevation of main valley floor.	Formed when main valley was glaciated and tributary was not, or where thin tributary glacier joined much thicker main glacier.
Bridalveil falls (vg, rarely cg)	High waterfalls over which stream drops from hanging valley into main valley.	See above.
Roche moutonnée (vg, cg) (fig. 6.4)	Ice-eroded bedrock boss, smooth on stoss side, irregular on lee side.	Abrasion on up-ice (stoss) side, crushing and plucking on down-ice (lee) side.

Figure 6.3. Glacial Landforms

II Depositional Features

	Topographic Description	How Formed
A. Glacial ice deposits (till) Ground moraine (t, vg, cg) (fig. 6.5)	Gently undulating blanket deposit—"swell and swale."	Subglacial deposition by live ice; or let down by melting of stagnant ice.
End moraine (fig. 6.6) (t, some sd, vg, cg)	Irregular ridge—"knob and kettle"; sometimes occurs in great loops.	A glacial dump at edge of live ice; trend perpendicular to direction of ice flow.

Three special kinds of end moraines: *terminal*—outermost end moraine of any glacial advance; *recessional*—any end moraine built by retreating glacier; *interlobate*—end moraine built between two lobes of ice.

	Topographic Description	How Formed
Drumlin (t, cg) (figs. 6.4, 6.7)	Smooth hill, inverted spoon shape, steeper on stoss side; not usually found singly.	Glacial debris, overridden by moving ice and shaped into teardrop form; long axis of drumlin parallel to direction of ice flow.

The following types of moraine accumulate on the surface of valley glaciers and may remain after ice has melted away; ablation moraines are also formed on continental glaciers.

	Topographic Description	How Formed
Lateral moraine (coarse t, vg)	Embankment along edge of glacier.	Accumulation of weathering debris (talus) along edge of glacier.
Medial moraine (t, vg)	Embankment down center of glacier.	Joining of two valley glaciers.
Ablation moraine (t, vg, cg)	Surface blanket of debris on glacier, especially near snout.	Melting and sublimation of ice; becomes part of ground moraine when ice is completely gone.

* (vg = valley glaciers; cg = continental glaciers; t = till; sd = stratified drift)

Figure 6.3. (Continued)

103

II Depositional Features (cont.)

	Topographic Description	How Formed
B. Water-laid deposits (stratified drift) The first three landforms are deposited by glacial meltwater within the confines of the glacier. Configuration of features may be determined by shape and position of ice.		
Esker (sd, cg, rarely vg)	Narrow, steep-sided, sinuous ridge.	Open crevasse or ice tunnel.
Kame (sd, vg, cg)	Irregularly shaped hill.	Hole in the ice or space between blocks.
Kame terrace (sd, vg, cg)	Terrace along valley side.	Space between glacier and valley wall.
Kettle or kettlehole, ice-block hole (vg, cg)	Undrained depression of irregular shape.	During deglaciation large blocks of ice are surrounded or buried by drift; when ice melts, slump or kettlehole remains.
The following two landforms are deposited by glacial meltwater outside the area of glacial ice. Overall shapes determined by meltwater activity and sediment load.		
Outwash plain (sd, cg) (*A pitted outwash plain* is pocked by numerous kettles.)	Smooth, broad, gently sloping plain, similar to coalescing alluvial fans.	Heavily loaded streams of meltwater, braided channels.
Valley train (sd, vg, cg)	Smooth, gently sloping deposit over valley floor.	Heavily loaded streams of meltwater occupying stream valley or former glacial trough.

Figure 6.3. (Concluded)

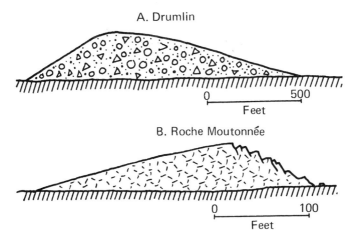

A. Drumlin

0 — 500
Feet

B. Roche Moutonnée

0 — 100
Feet

Figure 6.4. **A.** Cross-section of drumlin (made of till) deposited by an ice sheet moving from left to right. **B.** Profile of a glaciated bedrock mass in the shape of a roche moutonnée. The ice that eroded this landform came from the left.

water-laid deposits within its boundaries. Deglaciation by stagnation and breaking up is termed *downwastage*, while the melting back of the ice edge at the same time that ice is moving forward is called *backwastage*. A sequence of deposits consisting of compacted till with end moraines records the work of actively flowing, comparatively thick ice. In contrast, a sequence consisting of ablation till and ice-disintegration features, such as kames, eskers, and ice-block holes, is produced by ice that is thin, flowing very slowly or entirely stagnant (Flint, 1971).

GLACIER-MODIFIED LANDSCAPES

Each of the glaciated regions in the world has its own unique history and characteristic pattern of landforms producing a distinctive landscape. And yet similar groupings can be found, suggesting that fundamental relationships can be studied and associations made. Three associations of glacial processes, landforms, and landscapes are discussed here; namely, (1) alpine (valley) glaciers in extensive mountain areas of high elevation and high relief, (2) continental glaciation over lowland areas of low relief, and (3) continental glaciation over mountainous areas of moderate elevation and relief.

Alpine Glaciation

Alpine glaciers occupy and modify, by erosion, valleys already shaped by stream erosion and mass wasting. Active glaciers tend to increase the relief of mountains and valleys; sharpen the landforms, especially ridges and peaks; deepen and straighten the valleys; scour out undrained basins; and, because of unequal erosion and differential resistance, bring about the development of discordant valley junctions (hanging valleys) and waterfalls (*e.g.*, bridalveil falls). The jagged alpine skyline rises and dips through a succession of horns and cols. Just below lie the great snow-fields that feed the cirques at the head of the glacial troughs. The long, precipitous slopes of these immense features produce an effect of great openness and magnificence (Mount Rainier National Park, Washington, fig. 6.1).

Because of their relatively small size, the depositional features, both ice-deposited and water-deposited, of alpine glaciers seem inconspicuous in the total landscape. A moraine, for example, 150 feet high and perhaps 1,000 feet wide, that would be a landmark on the Illinois prairie, is lost in heavily forested mountain country where the drainage relief may be from 3,000 to 5,000 feet. Nevertheless, the capacity of valley glaciers to transport enormous quantities of material of all sizes is clearly evident. In addition to boulders, pebbles, and fine materials being carried be-neath or within the ice, debris from talus and avalanching rides down in lateral and medial moraines. After the ice has melted back, end mo-raines and recessional moraines may make cross-valley ridges on the floors of glacial troughs and, in some cases, dam streams to make lakes. Valley train deposits extend for many miles downstream from the ter-minus of a glacier. This deep valley fill may be cut into terraces that can be related to advances and retreats of the glaciers (fig. 3.6, p. 43).

Continental Glaciation in Areas of Low Relief

The Pleistocene ice sheets that advanced and retreated several times over central North America and central Europe covered regions that in preglacial times were topographically low as well as having low relief. The preglacial landscapes might be visualized as made up of low, roll-ing hills and well integrated, large-sized drainage systems, underlain by only moderately resistant rocks. What, if any, relationship did the pre-glacial landscapes have to those of the postglacial period?

The great volumes of glacial drift left by the ice indicate that the amount of erosion was considerable even though little evidence can be seen. Bedrock was scoured, and stream valleys were enlarged and deepened. In North America, the lowlands now occupied by the Great Lakes were probably major stream valleys. Data from geophysical prob-ing and well drilling have provided clues as to the locations of large preglacial river systems that formerly drained the area between Ohio and the Dakotas. Ancient river valleys, now filled with glacial drift, are valuable reservoirs of ground water.

Most of central and eastern Canada is a flat lowland with scoured resistant bedrock outcropping in many places. The landscape was scraped and smoothed by the continental ice sheets but not deeply eroded. South

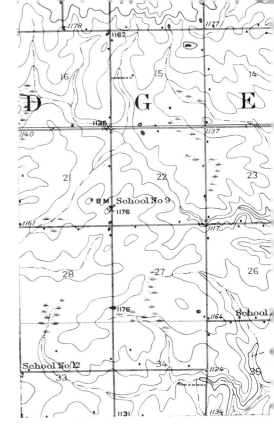

Figure 6.5. Boone, Iowa (1:62,500, C.I. 20 ft., 1948). The ground moraine on this upland has been slightly eroded. The drainage pattern is dendritic, but the numerous swamps indicate that the stream network has not achieved any semblance of integration. The streams are ungraded over most of the region, which has the characteristics of very early youth in the Davisian cycle. (Compare Fig. 5.3, p. 76, Casey, Illinois).

of the scoured area of Canada in a wide arc from the Great Lakes to the margin of glaciation, the ice deposited a thick blanket of drift, covering practically the entire land surface and constructing a new landscape. In Europe, a similar sequence of glaciation moved vast quantities of regolith and bedrock from northern Finland and Scandinavia down over the plains of central Europe.

The landforms and landscapes found in the central lowland areas of glacial deposition are chiefly morainal. *Swell and swale topography* dominates in the areas of ground moraine (Boone, Iowa, fig. 6.5). Here, drainage is poorly developed and relief is so low that the unevenness of the land surface is barely perceptible to the eye. Undrained depressions hold swamps or shallow lakes. Nearly every spring, large shallow pools (10 to 50 acres in size and a few inches deep) accumulate in the cornfields on the ground moraine surfaces of central Iowa and Illinois.

The landscapes of the end morainal belts are easily recognized by their *knob and kettle topography*. Local relief is not high; but the texture of the surface is such that one sees small hills, ridges, saddles, kettleholes (some occupied by small lakes), and many short, irregular slopes facing in all directions. The morainal belts form long, low ridges, perhaps 200 feet high and a mile or more in width (fig. 6.6).

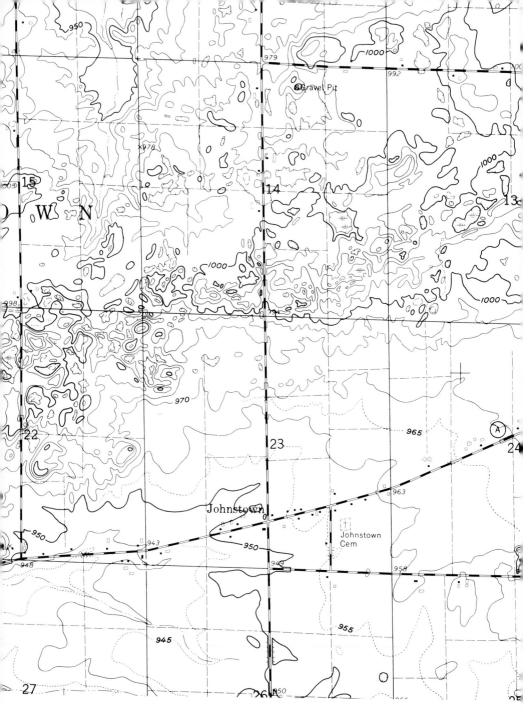

Figure 6.6. Avalon, Wisconsin (1:24,000, C.I. 10 ft., 1961). The south side of an end morainal ridge, showing knob and kettle topography; farther south, on ground moraine, is smoother swell and swale topography. Postglacial stream erosion has not significantly modified the glacial depositional topography.

Fields of drumlins, extending hundreds of square miles, dominate some areas of ground moraine. Examples can be seen north of Minneapolis, in central Wisconsin, and along the south side of Lake Ontario. As drumlins are oriented with the long axes roughly parallel to the direction of glacier movement, they give the landscape a definite grain (Palmyra, New York, fig. 6.7).

From the abundance of ice-deposited morainal features in the central lowlands and the scarcity of glacio-fluvial deposits (stratified drift), it appears that continental ice was active over most of its history and that it continued to carry debris up to the ice margins even while the edges were melting back. Backwastage was evidently dominant during deglaciation.

Glacial meltwater determined the location of many present-day streams. The valleys of the lower Missouri and Ohio Rivers, for example, were formed by streams draining along the edges of continental icecaps. Numerous south-trending rivers that cross the prairie states developed from meltwater flowing across fresh drift plains. These rivers still continue, although with lesser volumes, at the present time (see p. 59).

Continental Glaciation in Areas of Moderately High Relief

The highland areas of northwestern United States and northwestern Europe were covered over and eroded by the same continental ice sheets that spread over the lowlands during the Pleistocene. And yet the landscapes that resulted are strikingly different from those described in the previous two examples of alpine glaciation and glaciated areas of low relief.

Field data indicate that Pleistocene icecaps completely overrode the moderately high mountains of New England and New York in North America and the northern highlands of the British Isles in Europe. Indirect evidence, based on known measurements, suggests that in many areas the ice sheets were more than a mile in thickness, and possibly two or three miles thick in some locations. The magnitude of the icecaps can be visualized by thinking of the physical geography of northern New Hampshire and the adjoining parts of Quebec. The summit of Mt. Washington, New Hampshire, which was severely glaciated (along with the rest of the Presidential Range), is 6,288 feet above sea level. The St. Lawrence River valley, approximately 100 miles to the north, is at or slightly above sea level. Since the upper surface of the ice must have had a southward slope in order to pass over the top of the Presidential Range, the glacier must have been at least a mile and a half thick in southern Quebec.

The weight of such a mass of ice pushing itself over the land surface drastically altered the preglacial landscapes. Extensive abrading and smoothing of mountain tops and highlands produced "giant roches moutonnées" that make up a skyline of convexly rounded summits with severe scouring on their stoss (north) sides and deep plucking on the lee (south)

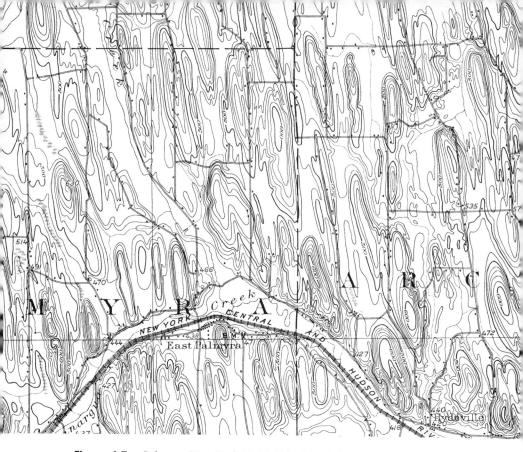

Figure 6.7. Palmyra, New York (1:62,500, C.I. 20 ft., 1921). The numerous hills of this area make up a drumlin field on a ground moraine surface. The wide flat valley toward the south was probably eroded by meltwater during deglaciation; hence the smaller stream now draining the area is misfit (Chapter 4).

sides. A climber going up any one of these steeper south slopes passes through great jumbles of ice-quarried blocks. A famous example of glacial plucking is the Old Man of the Mountains (Hawthorne's "Great Stone Face"), which looks majestically to the south from Cannon Mountain in New Hampshire. Because of intensive scouring and plucking, most of the upland surfaces in these glaciated regions have thin soil cover or none at all (Lovewell Mountain, New Hampshire, fig. 6.8).

Some interesting variations were caused by the orientation of the major preglacial landforms. Where a mountain range and valleys ran east and west, the mountain tops were obstacles and were severely abraded. However, in valleys oriented in approximately the same direction as that of ice flow, more ice passed through the valleys than over the ridge tops. Crawford Notch in the White Mountains of New Hampshire is typical of north-south valleys that were scoured into wide glacial troughs with steep, smooth walls. Where the major preglacial valleys were oriented transverse to the direction of the flow of ice, they became

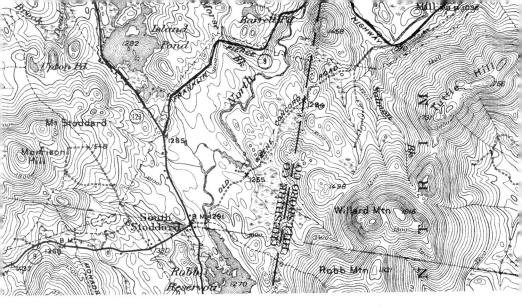

Figure 6.8. Lovewell Mt., New Hampshire (1:62,500, C.I. 20 ft., 1957). A landscape produced by continental glaciation on a region of moderate relief. The mountain tops are smoothed off and intervening lowlands are filled with glacial drift. The drainage pattern is disarranged and lakes and swamps are numerous.

choked with great volumes of drift that effectively disorganized the preglacial drainage.

From the appearance and nature of the deposits left in the preglacial valleys after the ice had melted, we can reconstruct a picture of the deglaciation process in areas of moderate relief. Hard, tight till (packed down underneath moving ice) is present but thinner and less extensive than on the broad plains of the central lowlands. In upland areas (on middle and lower slopes of mountains and in the high meadows) till is found plastered against bedrock. Overlying the till in some places and banked up on the mountainsides are stratified drift deposits. The presence of ice for some length of time was necessary for the deposition of these bedded accumulations of sand and gravel at the relatively high elevations. Stagnant ice masses provided watersheds and channels for aggrading glacial streams. Crenulated ice-contact slopes can be traced along many hillsides. On the lowlands below, glacio-fluvial streams have left multiple kame terraces, occasional eskers, lake deposits, and innumerable valley trains and outwash plains. Here and there, where immense blocks of ice were buried, gradual thawing over the centuries has left slumps and kettleholes. Drainage tends to be disorganized, with streams flowing through swamps and ponds, across bedrock ridges, and through postglacial valleys cut in drift. Lowlands are everywhere dotted with small lakes occupying kettleholes and uneven depressions in the drift. Larger lakes lie in valleys dammed by till and unevenly deposited outwash laid down over a bedrock surface (that initially had considerable relief).

Landscapes produced by continental ice sheets on regions of high relief are a mosaic of both erosional and depositional features, some much younger than others; whereas the glaciated landscapes of regions of low relief are, as we have seen, mainly depositional. Both geomorphic regions are being eroded by streams, so that the landscapes are, in a taxonomic sense, polygenetic. Where preglacial topography is still evident in the landscape, a region could also be referred to as multicyclical.

Periglacial Processes and Permafrost

In regions of year-round cold temperature, in the polar zones and high altitude areas, weathering and mass wasting are intensified (Embleton and King, 1968). The term *periglacial* is applied to these processes, the prefix "peri-" suggesting the environment around and beyond the edges of glaciers. Effects of periglacial processes are obvious and dominant in only a few places on the earth's surface; but in a minor, seasonal way, these processes operate nearly everywhere that freezing and thawing occur as part of the general erosional activities. In the less severe climate of humid temperate zones, water within a few feet of the surface freezes each winter and thaws out in spring. The depth of frozen ground may extend from several inches to several feet, depending on climatic conditions. Continued freezing and thawing of water in the regolith churns up material, breaks up particles and causes some differential sorting. In periglacial areas effects of these activities are multiplied. Sorting, shifting, wetting, and drying of the regolith may produce a patterned ground on which larger stones are concentrated in polygons or rings 30 to 50 feet in diameter and on slopes into long stripes and nets. On slopes, alternate freezing and thawing accelerates creep and solifluction (see p. 23).

In very cold parts of the world, where the average temperature stays below freezing the year round, water in the ground is permanently frozen, often to depths of several hundred feet. This *permafrost*, or permanently frozen ground, found mostly in the Canadian and Soviet Arctic, is believed to be a relic of colder Pleistocene temperatures. Some permafrost is slowly thawing out under warmer post-Pleistocene climates. Under special conditions, spectacular erosional processes may operate in a permafrost environment. Because meltwater and rain cannot infiltrate the ground, runoff is high, even in areas of low precipitation. When a few hot sunny summer days melt the permafrost near the surface, the regolith becomes a saturated mass of mud. On even a slight slope, the thawed material moves or flows downward into nearby valleys. Material disturbed and moved by frost action and melting may become distorted, folded, faulted, and thoroughly mixed, as the slope is denuded. As new, frozen regolith is uncovered and exposed to thawing, it, too, flows down the slope. This type of solifluction causes rapid erosion.

On relatively flat areas of permafrost, water collects in shallow *thaw lakes*. The warming effect of the water melts permafrost around and under the lake, causing the basin to grow in size. The picture of boggy, black-fly-infested "bush" typical of some polar areas in the Northern Hemisphere is due more to poor runoff and very low infiltration than to heavy precipitation. In terms of measurable rainfall (converted from snowfall), these Arctic areas are classified as semiarid or arid.

When man tries to live and work in permafrost areas, he may produce a variety of unwanted situations. A blacktop road or airport runway may collect enough of the sun's heat to melt the underlying permafrost and cause the roadway or the runway to collapse. Water supply, water or fuel transmission, and sewage disposal in permafrost areas present great difficulties. With the ecology of permafrost areas in such delicate balance, seemingly minor changes made by man may bring about significant or even disastrous alterations in the environment. When people live and develop resources in active periglacial or permafrost areas, special engineering and construction techniques and practices need to be carefully worked out (Haugen and Brown, 1970).

During the Pleistocene, when global weather patterns and wind circulations were different than at present, periglacial activities were much more extensive in a broad belt several hundred miles wide, south of the borders of continental glaciers in Europe and North America. A recent discovery of polygonal ground, stone rings, and stone stripes in the Appalachian Mountains of West Virginia and Virginia, well south of glaciated areas, indicates that colder Pleistocene climates probably produced these "fossil" periglacial features.

Wind action, in the past as well as the present, has been a significant aspect of periglacial climate. In the absence of vegetation, because of the cold temperatures, wind deflates the fine-sized sediments, which are transported and accumulate elsewhere as loess (p. 119). Periglacial winds may also build up sand dunes. The recognition and study of ancient dunes in northeastern Europe showed that Pleistocene wind patterns were different from those of the present. Wind-cut stones (ventifacts) and loess found in southern New England suggest that stronger winds operated during the retreat of Pleistocene ice than occur currently.

SUMMARY

Glaciers are formed by the compaction, crystallization, and freezing of snow. They flow plastically and by complex mechanisms of slippage. Rate and fluctuation of flow vary as a function of climate, altitude, and latitude. A cool, humid climate is optimal for glacier development. Glaciers erode by abrasion, plucking, and crushing. Under favorable conditions, quantities of regolith and rock are transported superglacially, englacially, and subglacially. Glaciers produce distinctive erosional and depositional landforms (fig. 6.3). Drift is the general term for all glacial deposits—till referring to those deposited by ice, and stratified drift to those deposited by meltwater. Alpine scenery is predominantly erosional.

Continental glaciation on areas of low relief smooths and then buries pre-existing topography as drift is deposited. In areas of moderate relief, uplands are eroded and lowlands are filled with drift.

Landscapes That Are Mainly the Result of Glaciation

Glacier landscapes: Antarctic icecap, Greenland icecap.

Landscapes developed by valley glaciers: Mountains with saw-toothed divides, cirques, glacial troughs (Swiss Alps; New Zealand Alps; Grand Tetons, Wyoming; Glacier National Park, Montana).

Landscapes developed by continental glaciers: Eroded bedrock plains (west, south, east of Hudson Bay); smoothed mountains with drift on slopes and in valleys (New Hampshire, Vermont, northeastern New York); swell and swale topography (young glacial plains of north central United States); knob and kettle topography (in belts across glacial plains, north central United States); kame and kettle topography in areas of low relief (eastern and southern New England); drumlin fields (central Minnesota, central Wisconsin); outwash plains (Cape Cod and southern Long Island).

Fiord landscapes: (eroded by valley and/or continental glaciers): Labrador; British Columbia; South Island, New Zealand; Norway.

Periglacial landscapes: Northern Siberia, Canada, and Alaska; at or above the snowline on the world's highest mountains.

Chapter 7

Wind, Waves,
Ground Water and Volcanism

Although running water in streams and rivers does most of the routine erosional and depositional work around the globe, several other geomorphic agents—wind, waves, ground water, and volcanic forces—usually get more publicity! All develop distinctive landscapes although their characteristics may be obscured by weathering and stream erosion. Geographically, these unique landforms are not widespread, but scientists and curious travelers who seek them out have long found them worthy of study as well as fascinating to observe.

WIND AS AN AGENT OF EROSION

Nearly always man is aware of the wind—both its presence and its absence. Occasionally the wind moves with great force—driving rain, breaking or uprooting trees, and even demolishing buildings. Because of our awareness of the great and variable power of the wind, we tend to overrate the amount of erosion done by its continuing blasts. Careful observation and study, however, have shown that wind is a comparatively inefficient agent of erosion and actually does little of the continual wearing away of the rocky earth and the transporting of sediments from one place to another.

The wind does serve as an important support for other agents. Within the hydrologic cycle, wind is the atmospheric mixer that moves great quantities of water (as water vapor) from the sea back over the land. Wind supplies the energy for most waves. Why it is, then, that wind in its normal activities does not erode effectively?

Moving air flows like moving water, but is considerably less dense. Wind cannot move material in suspension or by traction and saltation so effectively as water can. Nevertheless, where small particles are available, wind can pick them up, and carry them some distance high in the air. Silt-sized particles are moved along in suspension, while particles of sand size are swept along closer to the ground by traction or saltation. Because of this layering effect, most *wind abrasion* occurs within a few feet of the surface (Bagnold, 1941). An absence of grass or trees in open countryside allows a strong wind to get at particles, pick them up, or roll them along. This process is *deflation*. Conversely, anything that obstructs or slows down wind tends to cause deposition. Windbreaks, to be effective, need to be placed at right angles to the prevailing wind.

Humid regions and areas protected by vegetation are not entirely unaffected by wind erosion, but the effects of deflation and abrasion are more prevalent and more noticeable in arid and semiarid regions. Long continued deflation on an exposed area may produce circular depressions called *blowouts*. Those in western United States are usually referred to as "buffalo wallows." Pebbles, larger stones, and other resistant materials too firmly set or too heavy to be moved by the wind, remain in the bottom of the depressions. Where the effects of deflation are spread over a plain or flat basin area, a *desert pavement* of pebbles, or "lag gravels," may be formed. Playas (Chapter 4) in the arid regions are particularly subject to deflation of this kind. Runoff from desert rains brings in alluvium, some of which is subsequently borne off in suspension by wind. Eventually the action of wind or water or both may result in a desert pavement so tightly compacted that further deflation and erosion are inhibited.

In arid regions, as elsewhere, the effects of deflation are more significant than those of abrasion. However, saltating sand grains may abrade the surfaces of bedrock, boulders and pebbles, cutting and polishing them intensively. This "sandblast" action on exposed rock is limited to a few feet above the ground because sand grains do not rise very high in the air. The sand grains themselves tend to become rounded under long continued abrasion (Cooke and Warren, 1973).

Wind-deposited Landforms

Dunes, the most common wind-deposited features, are hills or ridges of well sorted sand. They may be *live* (active or moving) or *stabilized* by vegetation. Dunes are found anywhere that the factors necessary for their formation are (or were) present; namely, (1) a strong prevailing wind, (2) a source of sand, and (3) a place or cause for deposition. Because wind tends to pile up and move sand into characteristic shapes, dunes are classified as to the form they assume (fig. 7.1). *Transverse dunes,* formed in areas of abundant sand, are long, wavelike ridges, with furrows in between, that lie downwind of a linear sand source, such as a beach. Crests of the dunes are transverse (at right angles) to the wind direction (Porter, Indiana, fig. 7.2). Some transverse dunes, particularly along coasts, develop depressions or blowouts due to erosion by runoff or deflation. The blowouts become elongated or U-shaped as the dune moves downwind, and sand may be blown up into a hairpin-like ridge around the depression. Dunes of this shape are termed *parabolic. Longitudinal* or whaleback dunes are formed in localities where the wind velocity is fairly high, obstructions are few, and sand is not so abundant.

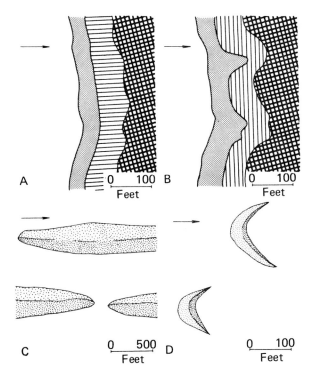

Figure 7.1. Diagrams of sand dunes, as seen from above: **A.** Transverse dune. Sand from the beach (at left) is accumulating in the dune (center) and migrating downward (toward the right). Compare Salmon Creek Beach in Figure 7.4. **B.** Parabolic or blowout dunes in a transverse dune complex. **C.** Longitudinal dunes (whaleback or sief). **D.** Barchan dunes.

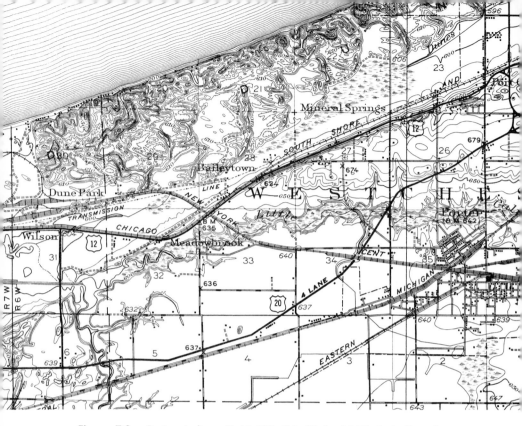

Figure 7.2. Porter, Indiana (1:62,500, C.I. 10 ft., 1940). In Indiana Dunes State Park a belt of sand dunes lies along the shore at the south end of Lake Michigan. With the beaches serving as sources of sand, the wind blowing off the lake has piled up a complex of transverse dunes. Local vegetation tends to halt the landward migration of the dune belt, although a few blowout or parabolic dunes have been pushed out from the main dune belt. South of the dunes is an area of ground moraine spotted with lake beds (Chapter 6).

The dune ridges run parallel to the wind direction. The great sand ridges of the central Australian desert are longitudinal dunes. *Barchans*, small, crescent-shaped dunes, are formed on flat surfaces and move downwind.

The windward sides of transverse, parabolic, and barchan dunes slope gently; the lee sides are steep, up to the angle of repose for dry sand. Wind sweeps sand grains up over the crest and drops them on the lee side, causing the dune to migrate. Some of the sand on barchan dunes is also blown around the sides, extending the arms in the downwind direction. On longitudinal dunes, sand is continually swept from the windward end along the sides and toward the leeward end.

Vegetation may partially or completely halt dune movement; but where vegetation is cut or worn away, the wind begins actively eroding again and either buries plants or prevents new plants from growing in. Some hills made of sand in characteristic dune shapes have become grassed over or even forested. Presumably these are ancient dunes formed in a drier or windier climate (perhaps during the Pleistocene) and later stabilized. The sand hills of northwestern Nebraska are stabilized dunes.

Loess

In several regions of the world—the upper Mississippi valley, eastern Washington, the Ukraine, Hungary, and north central China—a mantle of unconsolidated material, or *loess*, overlies other sediments and bedrock. (According to midwestern usage, *loess*, a colloquial German word for "soil," is pronounced "luss," to rime with "fuss"; alternate pronunciations are "lurse," to rime with "curse," and "lo-ess," similar to "Lois.") Loess, which consists of silt-sized particles, unstratified and poorly sorted, was first described and identified many years ago in China by Baron von Richtofen, a German geographer and explorer, who developed the desert theory for its origin. He postulated that the wind had blown for long periods of time from the arid interior (Mongolia) of the Asian continent eastward toward the more humid areas of central China. The loess particles were picked up by deflation, carried hundreds of miles in suspension, and then deposited in the region of the Yellow River (Hwang Ho) basin. (It is interesting to note that the Chinese term for loess translates as "yellow soil.")

When similar deposits were found to be widespread over much of the Middlewest, especially along the upper Mississippi and Missouri Rivers, it was at first thought that the loess had been wind borne from the "Great American Desert" (Colorado, Utah, Nevada, Arizona, New Mexico). But careful studies showed that the loess blanket was thickest near the major rivers and that it thinned away in an easterly direction. Moreover, the loess particles get smaller as one goes away from the rivers. It seems likely that the prevailing westerly winds of the central United States picked up silt-sized particles from deposits by rivers swollen with meltwater and loaded with glacial outwash. During low water stages, great sand and mud flats were exposed, providing immense sources of silt. On the bluffs and on the downwind side of the interfluves, vegetation slowed the wind and the loess was deposited. Along the eastern bank of the Missouri River in western Iowa, the loess is over 100 feet thick next to the river. Thirty miles to the east, loess is only a few feet thick. Similar loess deposits are found along the eastern side of the lower Mississippi River, especially near Vicksburg and Natchez.

Loess deposition has been observed in operation south of Fairbanks, Alaska, where the glacier-fed Tanana River serves as a source of wind-carried silt, which is collecting in the upland woods in a downwind direction. This explanation for the origin of loess, known as the glacial theory, appears to account for the American loess better than does the desert theory (p. 113).

WIND AS A SOURCE OF WAVE ENERGY

Most waves are formed by wind blowing over the surface of bodies of water—streams, lakes, and oceans. Since the size of the body of water limits the size of the wave that can be formed, big waves capable of doing major geologic work are found only on large lakes and the oceans.

In the formation of waves, energy is transferred from the moving air into wave shapes that move through the water. Waves are described in terms of height—the vertical distance between the top of the crest and the bottom of the trough—and length—the horizontal distance between wave crests. A wave period is the time required for a wave to advance the distance of one wave length. Small waves are a few feet high and a few tens of feet long. Great storm waves are a few tens of feet high and a few hundreds of feet long. Size of waves is a function of three key factors—the speed of the wind, the length of time that the wind blows in the same direction, and the fetch or distance of open water over which the wind is blowing. Without a fetch of several hundred miles, it is impossible for large waves to be produced. Once formed, big waves can travel thousands of miles into areas with very different wind patterns and arrive, eventually, off a coast as swells, quite unrelated to the local winds.

The ocean wave systems are related to (1) the planetary wind circulation, their primary source of energy; (2) the tides (resulting from gravitational attraction from moon, sun, and earth's rotation); and, very rarely, (3) to tectonic movements or displacements of the sea floor. But whatever the source of energy, the movement of waves around the earth are interrupted by the continental land masses. Exposed coasts, such as the western side of Ireland, the windward side of the Hawaiian Islands, the coasts of Peru and Chile, are the high energy coasts—the designation meaning that large waves in great numbers continually arrive along the shores. Protected shores, i.e., the west coast of Florida and the west coast of Japan, are coasts of low energy. In terms of geologic activity, wave erosion is most important and spectacular where large waves beat against a bold, unprotected coast.

Wave Action

In general, as they approach a shoaling coast, wave shapes are distorted by friction or drag along the bottom. The lower parts of the wave form are slowed to a point at which the still forward-moving top of the wave is left unsupported and hence falls or "breaks" forward. This is the action that produces breakers or combers. When waves break, a great deal of energy is available, and large masses of falling water slam against the bottom and shoot rapidly up the beach. Under the influence of gravity, the water then draws back down the beach slope in time to curl under the bottom of the next wave. Within the breakers, large volumes of water move rapidly in many directions under conditions of extreme turbulence.

Breaking waves are capable of a great deal of erosion and transportation. Materials within the breaker zone are swirled around, rolled, and bounced up and down the beach, banged into each other, and abraded by swirling sand grains. Here in the "wave-mill," rocks are broken up, rounded off, polished, and the pieces differentially sorted and transported. Some are taken up to the top of the beach; others are carried seaward along the bottom and deposited in quieter water offshore. Upshooting water undercuts cliffs; the pieces fall back, are broken up and carried away by the waves. During major storms, cliffs in unconsolidated material may be cut back many feet. Where waves beat against crystalline rocks, the rate of erosion is much slower; but joint blocks may be washed out and the bedrock scoured and abraded. Historic records for regions where large waves hit the coasts (e.g., the Channel coast of England, Helgoland, Boston Harbor, etc.) indicate that shorelines may be cut back hundreds of feet in a few centuries. Nix's Mate, a small island in Boston Harbor used in colonial times for the public hanging of pirates, has disappeared. At low tide all that can be seen is a "boulder pavement" on the shoals.

On most coastlines, elevation of the sea surface changes through two tidal cycles per day. In high and low latitudes tidal ranges are only a foot or two. In middle latitudes and in constricted waterways, the tidal range may be from 10 to 15 feet—in a few places exceeding 30 feet. In terms of wave erosion, the effect of tidal oscillation is in shifting the impact of the waves up and down the beach profile. Tides may be thought of as waves having relatively long periods (i.e., most commonly 6 to 12 hours). In the open sea the effects are not noticeable, but in the shore zones tides may conflict with or reinforce wind waves. When the *spring* (extra high) or *neap* (extra low) tides that occur each lunar month happen to coincide with strong onshore winds, both tides will be higher than usual. Combinations of high waves, spring tides, and onshore winds (as in gales lasting a day or more) are dreaded by inhabitants of coastline areas because when these unusually high water levels occur, the breaking waves and uprushing water can do extraordinary amounts of coastal erosion and damage to property.

Seismic sea waves (tsunamis) are destructive waves of great size and force that are generated not by wind but by submarine earthquakes or displacements of the ocean floor. Radiating outward from the source of seismic activity, tsunamis travel at high velocities (as much as 400 miles per hour in the Pacific, somewhat less in the Atlantic) all the way to the ocean boundaries. Tsunamis are not visible to ships on the open ocean, but as these waves hit a coastline, their tremendous force may send sea water in a series of inundations a hundred feet or more in height for some distance inland. Ships in harbors have been left high and dry in city streets after tsunamis have receded. The seismic waves occur rarely, but because of their catastrophic nature, they may do more erosive work on a coast than ordinary waves and storms could do in years or centuries.

Normal wave erosion is an oscillating process that transgresses and withdraws across a beach area (fig. 7.3). On many coasts, the longer, smaller waves of summer (a less stormy season) tend to pile up beach sand in low ridges or berms along the foreshore, giving it a convex profile and developing a full beach with plenty of sand. During winter storms, shorter, higher waves erode the foreshore, build up the cobble ridge on the backshore, and rake smaller sand particles out into offshore areas. Sometimes by spring, after a hard winter, a beach may have a flat or concave profile, quite deficient in sand, which is now stored in offshore bars below the water level, several hundred feet out from the breaker zone. In a geomorphic sense, the same stretch of shoreline may have a "summer beach" (fair weather) and a "winter beach" (stormy weather) that show changes in profile and sediment size seasonally.

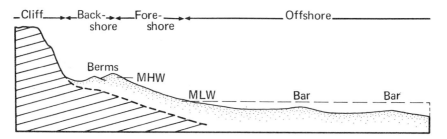

Figure 7.3.   Profile of a typical beach. The berms or beach ridges are built up by the waves. Offshore underwater bars are continually deposited, eroded, and redeposited by waves and currents. The erosional surface of the bedrock below the beach is a wave-cut bench. (MHW = mean high water; MLW = mean low water).

Sizes of material on a beach vary considerably across the beach profile. Coarse sediments and pebbles (called shingle) are found at the top of the beach and on the backshore, having been tossed there during recent storms. Coarse pebbles also may be found in the breaker zone, from which almost all the smaller-sized sediments are removed fairly rapidly. Sediment sizes may decrease up the beach slope to the top of the wave run and may also decrease seaward of the breaker zone. These zonations are temporary because the breaker zone moves up and down the beach with changing conditions of wind and tide. The most permanently located materials are those deposited on the backshore and in the deeper water offshore. Thus a beach can be described as the blanket of wave-washed sediments lying between these two limits and usually extending parallel to the coast. Typical beaches are located on *wave-cut benches* that were formed by big waves cutting away and removing bedrock or unconsolidated materials from cliffs and headlands. Shoreline studies have shown that most beach and cliff erosion occurs during severe storms. Between storms, the waves go on sorting, abrading, and shifting around the sediments of the beach.

Wave Refraction

On some shores waves come directly in on the beach with crests parallel to the trend of the beach. More frequently, the wave crests come in obliquely to a beach trend. The upwash (or swash) from the spilling breakers carries sediments obliquely up the slope; but the force of gravity causes the backwash along with sediments to return by the most direct route down the foreshore. Thus any material moved upward and then back has a net movement in a sideways direction. This process, called *beach drifting*, is an important element in the shape of many wave-built features. While the waves are arriving obliquely along the shore, water tends to pile up somewhat at the same time, developing a current flowing parallel to the shore. These longshore currents (that may also be generated by tidal conditions or other circumstances) carry sediments laterally along a shoreline, moving large volumes of sediments and building bars and other features, a process referred to as *littoral drifting* (Bodega Head, California, fig. 7.4).

On irregular coasts, the wave fronts are affected by the drag of shallow water in front of promontories, while on either side the crest of the wave continues with unabated speed until it, too, reaches the shallows. Each size of wave has a critical depth—4/3 of the wave height—that causes it to break. Thus ordinarily the wave breaks first in front of points or promontories and last at the head of the bay. In this way wave refraction brings about a concentration of wave energy against promontories and a decrease of wave energy—and hence less erosion—in the bays. Refracted waves hitting the points and headlands cause strong beach drifting and longshore currents that transport eroded debris from the headlands into the intervening bays, depositing sand and shingle on the bay heads. Theoretically, after the passage of enough time, an irregular coastline would be straightened, with the promontories cut back and the bays filled in. The geometric patterns of refracting waves can be observed and photographed from the air. Study of these patterns provides helpful information regarding the nature of the underwater topography close to the shoreline.

Wave Erosion Features

On mountain coastlines or wherever waves work against cliffs, terraces, or hills, erosional landforms evolve into spectacular scenery that is celebrated in art and literature and in family snapshot albums. One such coast lies along the southern Japanese islands, where *wave-cut benches* and *cliffs* in sedimentary rocks stand exposed to typhoons that come from

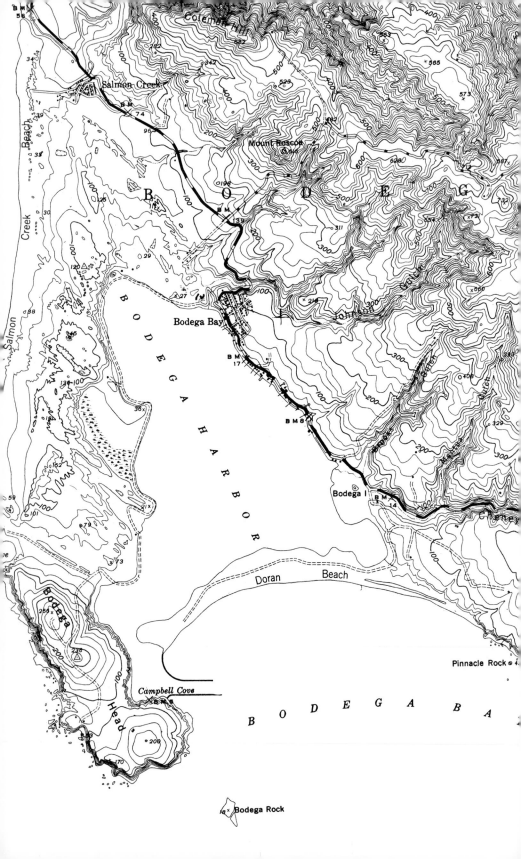

the south and central Pacific Ocean. On unprotected shores, waves work on zones of weakness along the joints and bedding of moderately resistant rocks. With waves eroding differentially, narrow points and inlets evolve, over several million years, into *arches, caves,* and *stacks* (cliff remnants). Examples can be seen on both sides of the English Channel and on the coasts of California, Oregon, and Washington.

Wave-built Features

Where sand and rock are abundant, waves transport material and build a variety of features, depending on the location of the sediment source (a headland, a stream delta, etc.), the configuration of the coast, and the underwater topography immediately offshore (which controls wave refraction). Where an excess of sand from shore drifting extends a beach laterally out into the water, a *spit* is formed. Usually spits tend to curve back toward land and are then described as *curved, recurved,* or *hooked spits. Arrows* are small spits that point seaward (Zenkovich, 1967). Sometimes spits from both sides of a bay join together, or nearly join, forming a *baymouth bar* or a *bay barrier.* A *tombolo* is a special kind of bar connecting an island with the mainland. *Cuspate bars* have a roughly triangular shape resulting from two sand bars building out from a shore in opposing directions and forming a point. The triangular lagoon left inside gradually fills in with sand ridges while more are built along the outer sides. In this way a *cuspate foreland* is formed, an example being Cape Canaveral, Florida. Not infrequently, offshore winds pile sand up on top of the bars and splits, creating dune complexes that further increase the elevations of the bars.

Along coasts where littoral drifting is predominantly in one direction, the bars and spits all extend in the downdrift direction. Where stream flow or tidal flow keeps a passage open through these bars, the openings (tidal inlets) are also displaced in the downdrift direction. If very strong tidal currents exist, they usually erode the end of a bar or spit and redeposit the material as a hook on the inside of the bar. In circumstances where drifting is very active, gaps in the bar, such as the mouths

Figure 7.4. Bodega Head, California (1:31,680, C.I. 25 ft., 1944). Doran Beach is a curved baymouth bar built across the mouth of Bodega Harbor. Beach drifting here is from the southeast in a westward direction (the bay outlet being on the western edge). The bar curved as it was built westward because it came under the shelter of Bodega Head, which provides protection from large waves coming from the west and also causes wave refraction. The two jetties at the mouth of the harbor are to prevent the closing of the harbor entrance by beach drifting. Between Bodega Head and the rocky mainland to the north lies Salmon Creek Beach. The irregular nature of the contours just in from the shore suggests the presence of a large area of coastal dunes.

of streams or harbor inlets, open and close frequently (fig. 7.4, Bodega Head, California).

CHANGES IN SEA LEVEL

Sea level elevation as a precise surface is difficult to determine because of variability of waves and tides. To calculate sea level, one first averages daily high tide with low tide. Daily values are, in turn, averaged to obtain a mean sea level. Anyone who observes coastal areas carefully can deduce that elevation of sea level has changed in relation to the elevation of the shore. Study of tide gage records kept for more than a century have shown that recognizable changes in sea level elevation have indeed occurred. Worldwide changes in the level of the oceans are referred to as *eustatic*. Deglaciation is an obvious cause of recent eustatic change; the melting of glaciers added more water to the oceans. Rate of eustatic rise has been measured as being a few inches per century.

Tectonic changes on the land also bring about changes in sea level. In the earth's mobile belts land may be rising or sinking as a result of mountain building or orogeny. A third type of change is the result of *isostatic rebound*. Under the weight of continental glaciers and the additional water of rising sea levels, the earth's crust was depressed; but after the ice and water receded, the crust began to rebound to its former elevation. Isostatic rebound is being measured at the head of the Gulf of Bothnia in the magnitude of half a foot to a foot and a half per century. Almost all coasts are undergoing changes of levels. In many places all three effects—eustatic change, tectonic change, and isostatic change—may be operating. Outside the glaciated areas largely eustatic and tectonic changes may be going on. Away from the mobile belts and areas of rapid sedimentation (such as that going on off the mouth of the Mississippi), chiefly eustatic changes would be affecting the elevations of land and sea. It is estimated that sea level was about 350 to 400 feet lower at the time of the maximum extent of Wisconsin glaciation. With the retreat of the glaciers, sea level rose rapidly (eustatic change) until about 6,000 years ago when the rate decreased sharply. Since then, sea level has risen the last few feet to its present position.

Recent changes in level can be recognized in some areas by distinctive features. Emergence of the land may bring up wave-cut cliffs, stacks, and wave-cut benches that now lie above the reach of the highest storm waves. Buried peat and occasional tree stumps that have been found below present sea level indicate submergence of the land, as do drowned stream valleys and glacial troughs (fiords). The erosional and depositional landforms of wave origin are probably destroyed by transgressing waves, and reports of such features being preserved following submergence are rare.

SHORELINE CLASSIFICATION

Geomorphologists, oceanographers, coastal engineers, and others have attempted to develop an ideal system for a simple, workable, genetic classification of the apparently endless variety of shoreline features existing in nature. A complicating effect of worldwide eustatic changes in sea level has been to give all shorelines some characteristics of submergence that overlap other features. Hence the problem of coping with all the significant variations has been a major difficulty in devising systems of shoreline classification.

Douglas Johnson's fourfold classification (1919) is convenient to use inasmuch as it is dependent principally on the configuration of a shoreline as shown on charts and maps. His classification can be summarized as follows:

Submergence. An irregular shoreline formed when a land area of moderate relief in a youthful or mature stage of erosion is submerged by an encroaching sea that laps up against spurs and flows up valleys, creating islands offshore and numerous wave-built features.

Emergence. An even shoreline, often with lagoon and barrier island, formed when shallow, smooth, offshore areas are elevated above sea level. The shoreline is regular because of the featureless nature of the recently elevated sea floor.

Neutral. A shoreline on which the configuration of the coast is not related to changes in the level of the sea—as, for example, where a lava flow, outwash plain, or fault has shaped the shoreline.

Compound. A shoreline with features indicative of both emergence and submergence.

The dominant geologic agent or process determines the classification of a particular shoreline in a system proposed by Francis Shepard (1963) that requires more knowledge about the local geologic history and the coastal environment. According to Shepard, *primary shorelines* are those on which the main features are the result of agents other than wave action; that is, rock structure and erosional history govern the shape of the shoreline. *Secondary shorelines* are those on which the features are formed principally by the action of the waves. Listed below are types of shorelines as classified by Shepard.

Primary (Youthful) Shorelines and Coasts

Land Erosion Coasts: Ria coasts (drowned river valleys), drowned glacial erosion coasts, drowned karst topography.

Subaerial Deposition Coasts: River deposition coasts, glacial deposition coasts, wind deposition coasts, landslide coasts.

Volcanic Coasts: Lava flow coasts, tepha coasts (where volcanic products are fragmental), volcanic collapse or explosion coasts.

Shaped by Diastrophic Movement: Fault coasts, fold coasts, sedimentary extrusions (salt domes, mud lumps).

Secondary Coasts

Wave Erosion Coasts: Wave-straightened cliffs, coasts made irregular by wave erosion.

Marine Depositional Coasts: Barrier coasts, cuspate forelands, beach plains, mud flats, salt marshes.

Coasts Built by Organisms: Coral reef coasts, serpulid reef coasts (cementing of worm tubes onto rocks or beaches), oyster reef coasts, mangrove coasts, and marsh grass coasts.

The suggestion that coasts be classified as retreating or advancing was put forth by H. Valentin (1954). According to his system, loss of coast can come about from erosion and/or submergence; growth is the result of emergence and/or deposition. Thus four "processes" may be in action. When one or more is dominant over the others, it (or they, as the case may be) determines the character of the coast.

A. L. Bloom (1965) has proposed a scheme for an explanatory description of coasts that involves three separate variables: erosion-deposition, emergence-submergence, and time. Bloom places these parameters on a three-axial diagram and shows how all types of coasts can be described as dynamic, changing landforms by means of this classification device.

The examples briefly outlined here of approaches to shoreline classification illustrate some of the ways of dealing with the problems involved. Johnson used the criteria of emergence and submergence plus time. Shepard and Valentin utilized the factors of erosion and deposition as well as emergence and submergence. Bloom combined all these variables in a diagrammatic approach illustrating the dynamic nature of shore features.

The categorizing of coasts as being "high energy" or "low energy," according to the potential power of waves that beat against them, has already been discussed. In a global and tectonic sense, the coasts of the world are also classified as "Atlantic types" or "Pacific types," depending on their relationships to major structural trends. On Atlantic-type coasts, the structural trends do not parallel the shoreline. As a result the coastal configurations tend to be irregular—as, for example, along the shores of northern New England, Newfoundland, Wales, and South Africa. On Pacific-type coasts, the major tectonic trends lie parallel to the coastline. This tends to produce an even coastline, as along western North and South America and Indonesia.

SUBMARINE FEATURES

Although some might question whether it is within the domain of the geomorphologist to extend his study of landforms and landscapes to the features of the ocean floors, many of the basic principles of subaerial geomorphology hold true in submarine environments. The major features of the ocean basins—like the dominant physical characteristics of the continents—are structurally controlled. The continental margins, the ocean floors, the great undersea mountain systems, and the abyssal trenches all have been formed by tectonic activity. In the shallower depths of the shelves and slopes around the continents are found most of the erosional and depositional features. Many of these areas have been charted and studied; but some of the scientific data are kept secret for economic, military, or political reasons.

A continental shelf is the zone that extends from the low-water line to the depth at which the ocean bottom drops off sharply at the continental slope. The average depth of the edge of the shelf (where it becomes the slope) is about 350 feet. Only 7.5 percent of the total ocean area is occupied by the continental shelves, but they equal 18 percent of the earth's land area (Emery, 1969). Pleistocene sediments, old beaches, terraces, former stream channels, drowned glacial troughs, roches moutonnées and other evidence of advances and retreats of continental glaciers have been charted on some areas of the North Atlantic continental shelf. A number of important fishing grounds (such as George's Banks about a hundred miles off Cape Cod) are largely of glacial depositional origin.

On the outer parts of the continental shelves are found the great submarine canyons—larger, on the average, than major stream valleys on land. Characteristically, these canyons have V-shaped cross-profiles, ungraded long profiles, and, with their tributaries, are arranged in dendritic drainage patterns. Despite their apparent resemblance to terrestrial stream systems, it is unrealistic to postulate the several thousand feet of raising and lowering of sea level which would have had to occur for submarine canyons to have been eroded on land and subsequently "drowned" even though some canyons appear to be extensions of drowned stream valleys. The turbidity currents (water holding quantities of sediment in suspension) that from time to time flush out loose sediments accumulating in the canyons may also have eroded them although other erosional processes, such as sediment creep, slumping, and bottom currents, have been observed at work in submarine canyons. Sediments accumulating in fan-shaped deposits (abyssal fans) outward from submarine canyons and as plains of deposition in structural basins appear similar to their counterparts on land and indicate that submarine processes are capable of large amounts of deposition. Gravitational forces also operate on the continental slopes and the deeper parts of the ocean causing slumping and sliding of sediments on both a large and small scale.

Seamounts

Extensive surveys of the ocean floors have revealed the presence of numerous *seamounts,* which are conical-shaped, steep-sided features un-

related to undersea mountain ranges. Many seamounts rise several thous-
and feet from the deep sea floor to within a few thousand feet of the
ocean surface. They are believed to be volcanic in origin. A curious feature of many are their flat tops—as if the seamounts had been truncated
by wave action at some time in the past. Subsequently the sea floor may
have subsided, lowering the seamounts to their present depths.

Coral Reefs

Corals (coelenterates) and, to a lesser extent, other marine organisms make up the faunal assemblages that construct extensive and in-
finitely varied features called *reefs*, or bioherms. Wherever conditions are
favorable, these colonial animals produce external skeletons in great pro-
fusion, creating—with the passage of time—the greatest organic structures that exist. Corals cannot live in water more than about 300 feet
deep because sunlight cannot penetrate to much greater depths; but in
warm clear water (free of sediment) where plenty of food is available
in the form of microscopic organisms, coral reefs may grow upward right
to the water surface. The resulting abrasive, stone-hard ridges have
plagued navigators of tropical waters for centuries.

Storm waves and the day-to-day action of surf erode the upper surfaces of coral reefs, breaking off loose pieces and endlessly grinding them.
Much of the reef debris falls or slides down the flanks of the ridge and
is gradually worked around by the waves, some of this material eventually becoming white coral sand on a nearby beach.

On reef tops living corals are intermixed with broken pieces of skele-
tal remains. Small channels and holes in the coral and knobs of various
sizes make the surface extremely irregular. Water from breaking waves
is thrown high in the air and falls back, pounding the reef with great
force. Boat handling near a coral reef is highly dangerous. By contrast,
the lagoons inside the reefs provide shelter from storms, and the quiet
waters are ideal for boating, fishing, and swimming.

Three general types of reefs are found (fig. 7.5). *Fringing reefs* grow
up against the rock of islands, extending the shoreline outward. *Barrier
reefs* are free standing and are usually separated from islands or mainland by a lagoon that may be too dirty or too deep to support coral growth.
If an island is conical in shape, a barrier reef may form an almost complete ring around the island. The Great Barrier Reef of Australia, on the
other hand, lies many miles off the coast and separates complex lagoons
from the open ocean with scarcely a break. An *atoll*, the third type of
reef, is an encircling feature lacking a central island. Examples are Eniwetok and Ulithi in the central Pacific, which were used as anchorages
by naval units and combat forces during World War II. A less common
type of reef structure occurs when great masses of coral grow up from
a shallow sea floor in the form of a head or rounded knob, called a *patch
reef*.

That an appropriate structural situation and favorable ecological conditions are essential for reef formation has long been known; but expla-

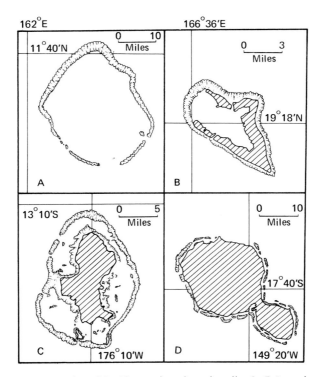

Figure 7.5. Examples of Pacific coral reefs and atolls: **A.** Eniwetok atoll and lagoon. **B.** Wake Island, a partial atoll with some central island remaining. **C.** Fringing reefs rim Wallis Islands; between the inner reefs and the outer ring of barrier reefs lies a wide lagoon. **D.** Tahiti, with both fringing and barrier reefs as well as lagoons.

nations for the occurrence of different types of reefs have not been completely satisfactory, although a variety of theories have been proposed. In general, the hypotheses attempt to explain how a platform for the reefs to grow on evolved and how changing levels brought about fringing reefs, barrier reefs, and atolls. Three types of explanations have been advanced: (1) The reefs were formed upon platforms that existed beneath the sea prior to the growth of the corals (antecedent platform theory); (2) corals grew, keeping up with the late glacial and postglacial rise in sea level (glacial control); (3) the rock foundations slowly subsided and the corals grew, staying at the level of the sea surface (subsidence theory). Inasmuch as none of these theories fits all types of reefs, it is probable that coral reef features were not all formed in the same way.

Drilling of cores and seismic exploration have indicated that a considerable thickness of coral exists on top of the volcanic rock of many atolls: at Funafuti about 1,000 feet; some 4,200 feet at Eniwetok; 3,900 feet at Bikini; and 3,000 to 6,000 feet at Kwajalein. These data suggest that the volcanoes and, presumably, the ocean floors have subsided 2,000 to 4,000 feet (Menard, 1964).

KARST TOPOGRAPHY AND EROSION BY SOLUTION

Over most of the earth's surface, ground water is the least obvious agent of erosion because so much of the evidence of its geologic work is hidden within the regolith and bedrock (p. 20). But in areas of adequate rainfall (either in the present or the past), underlain by rocks susceptible to solution, extensive and complex landscapes, called *karst topography*, may develop (Oolitic, Indiana, fig. 7.6).

Ground Water Movement

Ground water that penetrates regolith and bedrock, moving either downward or laterally, is capable of erosion, the principal mechanisms being solution and chemical precipitation (p. 15). Ground water usually moves too slowly to hold materials in suspension. The rate of movement and volume are mainly governed by the porosity and permeability of the rock materials (p. 12). If a bedrock is a carbonate that tends to be susceptible to chemical attack, the presence of thin bedding and extensive jointing sets up a situation favorable for weathering and erosion. Always seeking the path of least resistance, ground water works its way along joints and bedding planes and percolates among cracks and crevices, dissolving rock material and enlarging conduits. With larger passageways, more water can pass through, and it moves faster, ever increasing the amount of solution. When a fairly large cavity exists underground, the roof and walls may collapse, creating a sinkhole. When heavy rains fall above ground, the runoff, bearing sediments, washes into the depression, intensifying weathering and erosive activity. Where underground passageways and cavities are extensive, ground water is able to carry clay and other fine sediments along in suspension, sometimes depositing them far below the surface. Along the way, sediments tend to scour out and enlarge the conduits.

Erosional and Depositional Features Produced by Ground Water

The unique and varied features that are the result of weathering and solution by ground water range in size from small sinkholes to immense caverns but all come under the comprehensive term *karst*. Karst features develop in humid regions underlain by carbonate rocks (mainly limestones) and become extensive if the bedrock is well jointed and permeable and rainfall is moderate to abundant (figs. 7.6, 7.7).

The *caves* that form—especially those following intersecting planes and joints in the bedrock—may create a complex, three-dimensional network, linked by tunnels or a series of interconnecting rooms, some very large with high, domed ceilings. Some cave floors are flat; others slop-

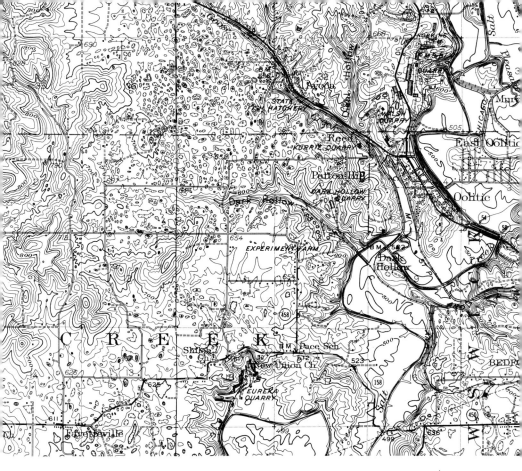

Figure 7.6. Oolitic, Indiana (1:62,500, C.I. 20 ft., 1935).Karst topography in this area is immature and solution by ground water is not far advanced. The small depressions on the uplands west of Salt Creek are sinkholes. The local limestone bedrock—notice the quarries along Salt Creek—is a building stone of high quality, famous for its oolitic texture. The incised meanders (Chapter 4) in the valley of Salt Creek appear asymmetrical and are probably the result of channel slipoff.

ing or step-like. Surface streams may flow into *sinkholes* or *swallowholes*, becoming "lost rivers" as they disappear into cave systems. Where erosion has been extensive, large areas of cave roofs may have fallen in, forming great depressions termed *uvalas*. Occasionally small sections of roof are left standing as *natural bridges*.

Where the bedrock is close to the surface, outcrops are softened, fluted, and pitted by rock weathering and solution. Eventually the entire surface may be carved up into vertically grooved blocks (*lapiés*) and honeycombed boulders as the carbonates are dissolved and washed away. Extreme karst topography may be so rough that passage across it, even on foot, is nearly impossible. Some of the most difficult terrain in the world to traverse—notably the Dinaric Alps in Yugoslavia—evolved by

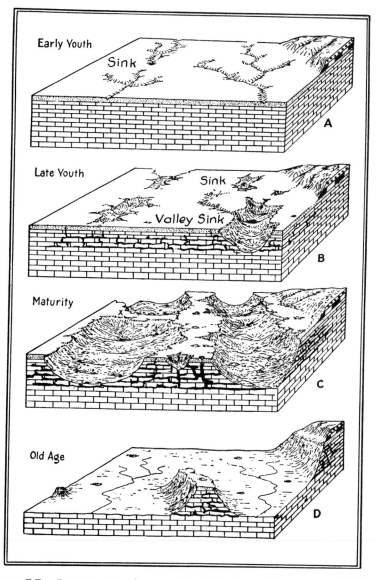

Figure 7.7. Four stages in the evolution of landscapes in an area underlain by limestone. Except for the earliest and latest stages surface drainage is is almost entirely absent. (After A. K. Lobeck 1939, **Geomorphology,** McGraw-Hill Book Co. Copyright © 1939.)

intensive karst action on folded sequences of thick carbonate rocks in an area of high relief.

In regions of flat-lying beds and low relief, as in central Florida, where the water table is high, sinkholes are filled with rising spring water (e.g., Silver Springs) and the landscape is dotted with small round lakes. Where collapsed blocks have partially filled the sinkholes, little circular groves of trees and bushes may grow that appear to be rising out of enormous sunken pots. Because of lowered water tables or tectonic changes (uplift or regional tilting), some karst areas—such as Carlsbad Caverns and some of the Virginia caves—have become fairly dry and open. In the Indiana-Kentucky karst region, where the limestone beds are horizontal or gently dipping (and the water table is lower than in Florida), sinkholes act like a regional sieve collecting most of the surface runoff, which then follows underground routes, sometimes for several miles before reappearing at the surface.

Weathering residues in Karst areas of low relief may form a characteristic red clay soil overlying the rock surface. *Terra rossa* ("red earth"), as this type of soil is called, is similar to the red lateritic soils of the tropics that are also formed as residual products of intensive weathering.

Depositional features formed by ground water include numerous cave forms made of dripstone (calcium carbonate precipitated out of solution). As ground water drips from cave ceilings, the water evaporates and leaves behind the mineral solute, which slowly piles up on the cave floor in thick conical forms called *stalagmites* and, at the same time, grows down from the ceiling as long, thin *stalactites.*

Where conditions are favorable, deposition of mineral material by ground water fills in around the grains of sedimentary rocks, effectively cementing them together. Similar secondary deposits are formed along cracks and on grain surfaces in crystalline rocks. When warm spring water comes to the surface, is cooled, and begins to evaporate, it deposits its dripstone load as travertine in the form of low mounds and terraces. Yellowstone Park is one of the best places to see these features. Chemically the material is similar to the "scale" that is formed in teakettles and plumbing by hard water.

ORIGIN OF LIMESTONE CAVES

Considerable controversy has existed as to whether ground water erodes mostly above or below the water table, the concomitant question being—How are caves and caverns formed in limestone? Some scientists hold the general idea that most of the geologic work of ground water is accomplished above the water table, the inference being that considerable mechanical erosion enlarges caverns begun by solution. Downward growth of cave systems thus is limited by base level of local streams (a factor which controls the water table) at any one stage. As streams cut deeper, new and lower cave systems are formed while dripstone accumulates in the drier, abandoned cavern passages. This hypothesis is called the *one-cycle theory.*

The *two-cycle* theory, on the other hand, suggests that water circulating and eroding (by solution) below the water table can result in extensive cave development. Proponents of this theory cite evidence that many cave systems have three-dimensional networks, blind galleries, ceiling pockets, and ungraded floors that could not have been formed by mechanical abrasion of running water. These investigators suggest that most of the cave erosion was accomplished by water circulating under hydrostatic pressure. At a later time, caves were drained of water as base level and the water table dropped, permitting air to circulate underground and bringing on the development of dripstone features. More recently, experts have suggested that cavern systems were developed by solution under pressure-flow conditions. Subsequently, streams diverted from the surface eroded well integrated cavern routes, further abrading the bedrock. As with many complex relationships in nature, it is probable that no single theory can adequately explain all the cave phenomena that exist.

VOLCANIC LANDFORMS

The process of volcanism produces geologic structures, landforms, and landscapes that are mainly constructional. Once formed they are subject to destruction by the usual forces of erosion, mainly by streams, but sometimes by glaciers, waves and winds (Cotton, 1952). The widely differing shapes and sizes of volcanoes found in nature are the result of different types of eruption and the chemical composition of the lavas. These controlling factors, therefore, form the basis for the usual classification of volcanic cones.

Volcanic Cones

When eruptions are relatively quiet and most of the erupted material is a lava of low viscosity, large broad mountains, called *shield volcanoes,* or lava domes, are built up. Sometimes these great cones are slightly convex in form, rather like the outer side of the shield of of a medieval warrior. The volcanic complex of Mauna Loa and Kilauea (both shield volcanoes) on the island of Hawaii are vents on top of a great mass of volcanic material that has been built up over 40,000 feet from the floor of the Pacific Ocean (Waipio, Hawaii, fig. 1.3, p. 7). The thousand-mile chain of volcanic islands making up the Hawaiian archipelago all have a similar history.

Stratovolcanoes, or composite cones, are the result of more violent eruptions, in which the ejecta include both fragmental material (pyroclastics) and high viscosity lavas. These cones are somewhat smaller and much steeper than those of the shield volcanoes. Eruptions of stratovolcanoes usually follow a sequence: first, explosion; next, eruption of ejecta; lastly, eruption of lava, which may pour out of fissures on the side of the cone rather than from the central crater. Thus by alternating layers of pyroclastics and lava, a cone is built up. These cones include some of the best known and most beautiful volcanoes of the world such as Fujiyama, Mount Vesuvius, and Mount Rainier (fig. 6.1, p. 95). The slopes of stratovolcanoes are slightly concave and the cones tend to be symmetrical. The elevations of the summits are usually between 10,000 and 15,000 feet.

In some types of volcanic eruptions, only ejecta and gases are thrown out. The ash, cinders, bombs, blocks, etc., fall back around the vent and accumulate as piles of debris, creating *cinder cones*. The sides are steep, usually at the maximum angle of repose for jagged material made up of a variety of sizes. Cinder cones are much smaller than stratovolcanoes and usually have a shorter eruptive life. Sunset Crater, near Flagstaff, Arizona, is a typical large cinder cone. The range in height of cinder cones is from a few hundred to a few thousand feet.

Craters and Calderas

Many volcanoes have depressions near their summits, formed by explosions and other eruptive activities. These smaller, funnel-shaped openings, mainly produced by explosive eruptions, are *craters*. Larger features, usually formed both by explosions and collapse, are called *calderas*. The distinction is mainly one of size. The use of genetic terminology here is of dubious value, inasmuch as the exact relationship between explosion and collapse is not completely understood. Collapse probably occurs after large amounts of ejecta have been thrown out by explosions, or after great volumes of lava have flowed out of the volcanic pipes and fissures. Crater Lake in Oregon is an example of a caldera formed by both explosion and collapse. Wizard Island is a more recent cone built within the caldera and represents a later resumption of activity.

Lava Plateaus

The largest and most significant examples of constructional landscapes to be found in the world are the lava plateaus and lava plains formed by the extrusion and hardening of eruptive material. The great lava plateaus of the Columbia River area in the Pacific Northwest, the Deccan plateau of west central India, the Yellowstone plateau of Wyoming, the Drakenberg plateau of South Africa, and most of the island of Iceland, all were built up, layer on layer, from great floods of low viscosity lava. Similar lavas, when flowing out at an elevation lower than the surrounding moutains, have built plains, as, for example, the Snake River plains in Idaho. Whether as plain or plateau, these masses of lava cover thousands of square miles and are often several thousand feet thick.

Features of Lava Flows and Volcanoes

Tumuli are expanding domes on the surface of lava, often with cracks in them through which liquid lava has flowed out. "Squeeze-ups" (as the name implies) are formed by the eruption of new lava through cracks in hardened lava. *Spatter cones* and mounds appear in infinite variety where small eruptions occur (Craters of the Moon, Idaho, fig. 7.8). Lava tunnels and lava caves, typical of low viscosity flows, are the result of hardening on the surface while underneath the still liquid lava was draining away.

The erosion of volcanoes has produced striking topographic forms. Solidified conduits and filled-in fissures tend to be more resistant to erosion than the pyroclastics and lavas of the cones. The hard necks and dikes remain as spires and sharp ridges long after softer materials have eroded away. Devil's Tower in Wyoming is an eroded volcanic neck. Ship Rock, near Farmington, New Mexico, is an erosion remnant composed of several radiating fissure fillings. Dikes and sills, because of differential resistance to erosion, may develop into hogback-like ridges or, in rare cases, as elongated trenches if the volcanic rock is less resistant than the surrounding country rock (p. 64).

SUMMARY

Wind and Waves

The wind and wind-driven water waves act as agents of erosion in accordance with principles developed from the study of hydraulic processes (Chapter 3). Wind erodes wherever it blows steadily or strongly enough to pick up material. Deposition occurs when wind slows down. Waves erode and transport earth materials and construct a variety of features. Wind is a minor agent of erosion, but waves, under favorable conditions, can do great amounts of geologic work. Erosional landforms produced by wind are blowouts and desert pavements. Depositional features are dunes of all types and loess blankets. Wave action produces such erosional landforms as wave-cut benches, cliffs notched at the base,

Figure 7.8. Craters of the Moon National Monument, Idaho (1:31,250, C.I. 10 ft., 1940). Recent volcanic activity, some of which occurred less than a thousand years ago, has created most of the constructional features of this landscape. The northwest-southeast arrangement of the cones suggests the location of a fissure zone, probable source of the eruptive material. The zigzag pattern of the contour lines drawn across the surface of Big Crater Flow gives evidence of the extreme roughness of the topography. The absence of stream valleys is due not only to the semiarid climate of this part of Idaho but also to the extreme youth and the high infiltration rates of the porous volcanic rocks.

stacks, sea caves and arches, and inlets through beaches. Depositional landforms include beaches, spits, tombolos, cuspate forelands, bars, barrier islands, and berms.

Arid Region Landscapes Produced by Action of Wind and Water

Sandy deserts (the erg of the Sahara).
Stony deserts (Atacama, Kalahari, Australian gibber plains).
Basin and range topography (Utah, Nevada).
Fans and playas (Mohave Desert, California; Lake Eyre basin in Australia).

Landscapes Modified by Wave Action

Long straight shorelines with lagoons and barrier islands lying offshore from areas of low relief (Texas coast).

Cuspate bars and forelands along and offshore from areas of low relief (Atlantic coast from Cape Henry, Virginia, southward to Charleston, South Carolina).

Deeply embayed shorelines on areas of moderate relief (Maine).

Irregular shorelines with beaches and wave-built features between rocky headlands (Oregon, New Hampshire, Massachusetts).

Wide, continuous beaches in front of cliffs cut in areas of moderate relief, usually on nonresistant regolith and bedrock (Outer Cape Cod, Massachusetts; Oregon and Washington).

Beach and swamp deposits where rapid alluviation is occurring (southwestern Louisiana).

Fringing and barrier reefs (Florida Keys, Bermuda).

Submarine Features and Landscapes

Erosion is minor and deposition is significant in the development of undersea topography, which is mainly controlled by the structural framework of the major ridges, deeps, and trenches. Some of the features that have been charted and studied are submarine canyons; tidal channels; glacial forms, both erosional and depositional; seamounts, coral reefs (bioherms) including patch, fringing, and barrier reefs and atolls; continental shelves and slopes; ocean basins; mountain ranges; tectonic deeps.

Ground Water

Ground water percolates within the earth's crust dissolving and transporting rock and earth materials. Some erosional landforms produced by ground water are sinkholes, swallowholes, uvalas, caves, caverns, natural tunnels, and bridges. Depositional features include terraces of sinter or travertine and accumulations of dripstone.

Landscapes Produced Mainly by Ground Water Activity

Karst topography in areas of high relief (Dinaric Alps, Yugoslavia; Carlsbad Caverns, New Mexico).

Karst topography in areas of flat-lying rocks (Indiana and Kentucky, including Mammoth Cave; north central Florida; Yucatan peninsula, Mexico).

Volcanism

Volcanic activity is basically constructional and produces such landforms as cones, craters, calderas, lava flows, volcanic necks, and dike ridges.

Volcanic Landscapes

Complex assemblages of volcanic cones (Hawaiian Islands).

Lava plateaus (Columbia River plateau in Washington and Oregon; Yellowstone plateau, Yellowstone National Park).

Lava plains (Craters of the Moon National Monument, Idaho).

Blankets of ash (near Rotorua, North Island, New Zealand).

Recapitulation

Individual topographic features are called landforms. Combinations of landforms make up landscapes.

The features of the earth's surface are shaped by geologic agents: running water, glacial ice, ground water, wind, and wind-driven water waves. Each of these agents picks up, transports, and deposits material. Tectonic and volcanic forces also contribute to the shaping of landforms and landscapes.

Most terrain features are formed by erosion or deposition or by a combination of both. An explanation for the origin of a specific feature lies in the nature of a particular geologic process or processes.

Features change with the passage of time. Both erosional and depositional processes produce sequential forms.

A particular process or combination of processes and the rates of erosion and deposition are determined mainly by climatic factors.

Tectonic forces may slow down or accelerate erosional and depositional processes.

Features produced by a combination of several erosional and depositional processes are polygenetic.

Features shaped by agents working in more than one sequence are multicyclical.

Example: A student living in southeastern Iowa (dissected drift plains) finds that the landscape was formed by the following combination of processes: glacial deposition; glacial and postglacial stream erosion and deposition; wind erosion and deposition (loess); continuous weathering, mass wasting and ground water solution in a humid temperate climate. The flat-lying buried bedrock has had little effect on the landforms and landscapes except in major stream valleys.

Bibliography

Expanded definitions of technical terms can be found in the *Glossary of Geology* (Gary *et al.*, 1972). For succinct discussions of topics, refer to the *Encyclopedia of Geomorphology* (Fairbridge, 1969).

BAGNOLD, R. A. (1954), *The physics of blown sand and desert dunes*, Methuen, London, 265 pp.

BLOOM, A. L. (1965), The explanatory description of coasts, *Zeitschrift fur Geomorph.*, N. F. 9, pp. 422-436.

BRYAN, K. (1940), The retreat of slopes, in O. D. von Engeln (ed.), *Sym-Am. Geog.*, v. 30, pp. 254-268.

CARSON, M. A., and KIRKBY, M. J. (1972), *Hillslope form and process*, Cambridge Univ. Press, London, 475 pp.

CHORLEY, R. J., DUNN, A. J., and BECKINSALE, R. P. (1964), *The history of the study of landforms*, v. 1, *Geomorphology before Davis*, Methuen, London, 678 pp.

CHORLEY, R. J., BECKINSALE, R. P., and DUNN, A. J. (1973), *The history of the study of landforms*, v. 2, *The life and work of William Morris Davis*, Methuen, London, 874 pp.

COOKE, R. U. and WARREN, A. (1973), *Geomorphology in deserts*, Univ. California Press, 374 pp.

COTTON, C. A. (1947), *Climatic accidents in landscape making*, Whitcombe and Tombs, Christchurch, 354 pp. (facsimile ed. 1969, Hafner Publishing Co.).

――― (1952), *Volcanoes as landscape forms*, Whitcombe and Tombs, Christchurch, 416 pp. (facsimile ed. 1969, Hafner Publishing Co.).

――― (1958), *Geomorphology* (7th ed.), Whitcombe and Tombs, Christchurch, 505 pp.

CURRAN, H. A.; JUSTUS, P. S.; PERDEW, E. L.; and PROTHERO, M. B. (1974), *Atlas of landforms* (2d ed.) John Wiley and Sons, 140 pp.

DAVIS, W. M. (1899), The geographical cycle, *Geog. Jour.*, v. 14, pp. 481-504. (also in *Geographical Essays*, reprinted by Dover Publications, 1957).

――― (1905), The geographical cycle in an arid climate, *Jour. Geol.*, v. 13, pp. 381-407. (also in *Geographical Essays*, reprinted by Dover Publications, 1957).

――― (1932), Piedmont benchlands and Primärrümpfe, *Bull. Geol. Soc. Am.*, v. 43, pp. 399-440.

DURY, G. H. (1960), Misfit streams, problems in interpretation, discharge, and distribution, *Geog. Rev.*, v. 50, pp. 219-242.

——— (1969), *Perspectives on geomorphic processes*, Assoc. Am. Geog. Resource paper no. 3, 56 pp.

DYSON, J. L. (1962), *The world of ice*, Alfred A. Knopf, 292 pp.

EMBLETON, C., and KING, C. A. M. (1968), *Glacial and periglacial geomorphology*, St. Martin's Press, 608 pp.

EMERY, K. O. (1969), The continental shelves, *Scientific American*, v. 221, Sept., pp. 106-122.

VON ENGLN, O. D. (ed.) (1940), *Symposium, Walther Penck's contributions to geomorphology*, Annals Assoc. Am. Geog., v. 30, pp. 219-284.

——— (1942), *Geomorphology*, Macmillan, 655 pp.

FAIRBRIDGE, R. W. (ed.) (1968), *The encyclopedia of geomorphology*, Reinhold, 1295 pp.

FENNEMAN, N. M. (1931), *Physiography of western United States*, McGraw-Hill, 534 pp.

——— (1938), *Physiography of eastern United States*, McGraw-Hill, 714 pp.

FLINT, R. F. (1971), *Glacial and quaternary geology*, John Wiley and Sons, 892 pp.

GARNER, H. F. (1974), *Origin of landscapes*, Oxford Univ. Press, 734 pp.

GARY, M., MCAFEE, R., and WOLF, C. L. (1972), *Glossary of geology*, American Geological Institute, 856 pp.

GILBERT, G. K. (1877), *Geology of the Henry Mountains*, U.S. Geog. and Geol. survey of the Rocky Mountain Region (Powell), 160 pp.

GOLDICH, S. S. (1938), A study in rock weathering, *Jour. Geol.*, v. 46, pp. 17-58.

HACK, J. T. (1960), Interpretation of erosional topography in humid temperate regions, *Am. Jour. Sci.*, v. 258A, pp. 80-97.

HAMMOND, E. H. (1954), An objective approach to the description of terrain (abs.), *Annals Assoc. Am. Geog.*, v. 44, p. 210.

HARBAUGH, J. W., and BONHAM-CARTER, G. (1970), *Computer simulation in geology*, John Wiley and Sons, 575 pp.

HAUGEN, R. K., and BROWN, JERRY (1971), Natural and man-induced disturbances of permafrost terrane, in D. R. Coates (ed.) *Environmental Geomorphology*, publications in geomorphology, State Univ. New York, Binghamton, 262 pp.

HEEZEN, B. C., and THARP, M. (1958), *Physiographic diagram of the North Atlantic ocean*, Geol. Soc. Am. (Similar diagrams of the South Atlantic and Indian oceans were published in 1962 and 1964 respectively.)

HORTON, R. E. (1945), Erosional development of streams and their drainage basins; hydrophysical approach to quantitative morphology, *Bull. Geol. Soc. Am.*, v. 56, pp. 275-370.

HUNT, C. B. (1967), *Physiography of the United States*, W. H. Freeman, 480 pp.

JOHNSON, ARTHUR (1960), Variation in surface elevation of the Nisqually glacier, Mount Rainier, Washington, *Intern. Assoc. Sci. Hydrol. Bull.*, v. 19, pp. 54-60.

JOHNSON, D. W. (1919), *Shore processes and shoreline development*, John Wiley and Sons, 584 pp. (Facsimile ed. 1965, Hafner Publications).

KELLER, W. D. (1957), *The principles of chemical weathering*, Lucas Bros., Columbia, Missouri, 111 pp.

KESSELI, J. E. (1954), A geomorphology suited to the needs of geographers (abs.), *Annals Assoc. Am. Geog.*, v. 44, pp. 220-221.

KING, L. C. (1953), Canons of landscape evolution, *Bull. Geol. Soc. Am.*, v. 64, pp. 721-752.

—— (1967), *The morphology of the earth*, 2d ed., Hafner Publishing, 699 pp.

LEOPOLD, L. B., WOLMAN, M. G., and MILLER, J. P. (1964), *Fluvial processes in geomorphology*, W. H. Freeman, 522 pp.

MACKIN, J. H. (1948), Concept of the graded river, *Bull. Geol. Soc. Am.*, v. 59, pp. 463-512.

MEIER, M. F. (1964), Ice and glaciers, *Handbook of applied hydrology* (V. T. Chow, ed.), McGraw-Hill, sec. 16, pp. 1-32.

MENARD, H. W. (1964), *Marine geology of the Pacific*, McGraw-Hill, 271 pp.

MORISAWA, M. E. (1959), *Relation of quantitative geomorphology to stream flow in representative watersheds of the Appalachian Plateau Province*, Columbia Univ. Dept. Geol. Tech. Rept. no. 20, Contract N60NR 271-30, 94 pp.

—— (1968), *Streams, their dynamics and morphology*, McGraw-Hill, 175 pp.

MORISON, E. E. (1966), *Men, machines, and modern times*, M.I.T. Press, Cambridge, Massachusetts, 235 pp.

MURPHY, RICHARD E. (1967), A spatial classification of landforms based on both genetic and empirical factors—a revision, *Annals Assoc. Am. Geog.*, v. 57, pp. 185-186.

—— (1968), Landforms of the world, Map supp. no. 9, *Annals Assoc. Am. Geog.*, v. 58.

OLLIER, C. D. (1969), *Weathering*, American Elsevier Publishing, 304 pp.

PENCK, WALTHER (1924), *Die Morphologische Analyse: Ein Kapitel der physikalischen Geologie*, Geographische Abhandlungen, J. Engelhorns, Stuttgart, 283 pp.

—— (1953), *Morphological analysis of landforms*, trans. by Czech, H. and Boswell, K. C., St. Martin's Press, 429 pp.

PLAYFAIR, J. (1802), *Illustrations of the Huttonian theory of the earth*, Wm. Creech, Edinburgh, 528 pp. (Facsimile ed., 1956, Univ. Illinois Press).

REEVES, R. G. (ed.) (1975), *Manual of remote sensing*, Am. Soc. Photogrammetry (in press).

REICHE, PARRY (1950), *A survey of weathering processes and products*, (rev. ed.) Publications in geol., no. 3, New Mexico, Albuquerque, 95 pp.

RUBEY, W. W. (1938), The force required to move particles on a stream bed, *U.S. Geol. Survey prof. paper* 189E, pp. 121-141.

RUHE, R. V. (1975), *Geomorphology*, Houghton Mifflin, 246 pp.

SCHUMM, S. A. (ed.) (1972), *River morphology*, Dowden, Hutchinson, and Ross, Stroudsburg, Pa., 429 pp.

—— and MOSLEY, M. P. (eds.) (1973), *Slope morphology*, Dowden, Hutchinson, and Ross, Stroudsburg, Pa., 454 pp.

SHARP, R. P. (1960), *Glaciers*, Univ. Oregon Press, Eugene, 78 pp.

SHEPARD, F. P. (1963), *Submarine geology* (2nd ed.), Harper and Row, 557 pp.

SHREVE, R. L. (1968), *The Blackhawk landslide*, Geol. Soc. Am. Spec. paper 108, 47 pp.

SIMON, MARTIN (1962), The morphological analysis of landforms: a new review of the work of Walther Penck. *Trans. of the Inst. of Brit. Geog.*, no. 30, pp. 1-14.

STRAHLER, A. N. (1950), Equilibrium theory of erosional slopes approached by frequency distribution analysis, *Am. Jour. Sci.*, v. 248, pp. 673-696, 800-814.

―― (1952), Dynamic basis of geomorphology, *Bull. Geol. Soc. Am.*, v. 63, pp. 923-938.

―― (1954), Empirical and explanatory methods in physical geography, *Professional Geographer*, v. 6, no. 1, pp. 7-8.

―― (1964), Quantitative geomorphology of drainage basins and channel networks, *Handbook of applied hydrology* (V. T. Chow, ed.), McGraw-Hill, sec. 4, pp. 39-76.

―― (1969), *Physical geography* (3rd ed.), John Wiley and Sons, 733 pp.

―― (1973), *Introduction to physical geography*, (3d ed.), John Wiley and Sons, 455 pp.

SUNDBORG, A. (1956), The river Klaralven: a study of fluvial processes, *Geogra. Annaler*, v. 38, pp. 125-316.

THOMAS, M. F. (1974), *Tropical geomorphology*, Halsted Press (John Wiley), 332 pp.

THORNBURY, W. D. (1965), *Regional geomorphology of the United States*, John Wiley and Sons, 609 pp.

―― (1969), *Principles of geomorphology* (2nd ed.), John Wiley and Sons, 594 pp.

TUAN, YI-FU (1958), The misleading antitheses of Penckian and Davisian concepts of slope retreat in waning development, *Proc. Indiana Acad. Sci.*, v. 67, pp. 212-214.

UPTON, W. B., JR. (1970), *Landforms and topographic maps*, John Wiley and Sons, 134 pp.

VALENTIN, H. (1954), *Die küsten der erde* (2nd ed.), Geographische-Kartographische Anstalt, Gotha, 118 pp.

WASHBURN, A. L. (1965), Geomorphic and vegetational studies in the Mesters Vig district, Northeast Greenland—General Introduction, *Medd. om Grønland*, v. 166, no. 1, 60 pp.

―― (1967), Instrumental observations of mass wasting in the Mesters Vig district, Northeast Greenland, *Medd. om Grønland*, v. 166, no. 4, 296 pp.

WAY, D. S. (1973), *Terrain analysis, a guide to site selection using aerial photographic interpretation*. Dowden, Hutchinson, and Ross, Stroudsburg, Pa., 392 pp.

WOLMAN, M. G., and MILLER, J. P. (1960), Magnitude and frequency of forces in geomorphic processes, *Jour. Geol.*, v. 68, pp. 54-74.

ZAKRZEWSKA, B. (1967), Review article: Trends and methods in landform geography, *Annals Assoc. Am. Geog.*, v. 57, pp. 128-165.

ZEIGLER, J. M., HAYES, C. R., and TUTTLE, S. D. (1959), Beach changes during storms on outer Cape Cod, *Jour. Geol.*, v. 67, pp. 318-336.

ZENKOVICH, V. P. (1967), *Processes of coastal development* (trans. by D. G. Fry; J. A. Steers and C. A. M. King, eds.), Interscience Publishers, 738 pp.

Index